'An incredibly well researched and entertaining book. Kasket has the ability to illustrate complex research areas through engaging, human stories that often reach all the way into your heart. As a recent parent, this book really made me reevaluate the way I use technology around my family. I do research on the impact of digital technology for a living. Even so, this is undoubtedly one of the most thought-provoking books I've read this year.'

– Carl Öhman, author of *The Afterlife of Data*

'A critical reminder that, at every stage of life, we get to choose our relationship with technology – and our choices shape our humanity. Kasket's book is a roadmap filled with generous possibilities.'

– Luke Burgis, author of *Wanting*

'Digital technologies aren't just transforming every area of life. They're transforming us, and this book, examining the psychological, social and technological intersections of this transformation, couldn't be more timely.'

– Catherine Meyer, author of *Good Grief*

Reset

**Rethinking your digital world
for a happier life**

Elaine Kasket

Elliott&Thompson

First published 2023 as
Reboot: Reclaiming Your Life in a Tech-Obsessed World
by Elliott and Thompson Limited
2 John Street
London WC1N 2ES
www.eandtbooks.com

This paperback edition published in 2024

ISBN: 978-1-78396-783-4

9 8 7 6 5 4 3 2 1

A catalogue record for this book is available from
the British Library.

Typesetting: Marie Doherty
Printed by CPI Group (UK) Ltd, Croydon, CR0 4YY

To the writers of the London Writers' Salon:
all those little boxes on the screen,
and in each one there was a gift.

I am a lost person. I wouldn't write books if I wasn't lost. I wouldn't write anything at all if I wasn't in search of paradise, and I wouldn't be in search of paradise if I didn't need it; if I didn't think I would be less lost if I were to find it. So I write to find it . . . but no, not that either, because I am nearing middle-age now and I know there is nothing to find. . . . There is no paradise out there, so I write to create my paradise on paper or on this blank, flat screen, but something in me always sabotages it and turns it dark. So then I write to reorder the world so that paradise might look possible again even for a moment, for someone.

– Paul Kingsnorth, *Savage Gods*

Contents

Introduction

At university, as virtually all budding psychologists do, I studied a man who's consistently listed among the top ten most influential psychological theorists of all time: Erik Erikson.[1] When we use phrases such as identity crisis, arrested development and midlife crisis, we're channelling him. His predecessor Sigmund Freud said that what happens in early childhood decides who we become,[2] but that didn't ring true for Erikson – Freud's theory of personality development was too inflexible, too deterministic, and didn't capture how our socially embedded selves keep evolving as long as we exist.

So, from the 1950s until his death in 1994, Erikson developed an enduringly popular model of how biological, psychological and social forces continually shape our identities. If you search for 'developmental stages' online, you'll likely find his model first. Breaking the life course into eight sequential phases, from birth to death, Erikson described how people encounter a key turning point at each stage. He expressed these as tensions between opposing forces, one positive and one negative: Trust vs Mistrust for infants; Identity vs Role Confusion for teenagers; Intimacy vs Isolation for young adulthood. An individual who's able to adopt helpful virtues such as hope, purpose, competency and love can better navigate and resolve each crisis at it comes, enabling one to move forward with greater psychological health and a stronger sense of self.

I know this because I wrote an essay about Erikson for one of my first classes in psychology, over three decades ago. I can't imagine how I did it, by which I mean I can barely remember how research was possible in the late 1980s. I can picture *where* I did it: the library, because there wasn't any other option for accessing information. Card catalogues were my search engine, and my sense memories of them revolve around touch: the handles of the draw-ers, the striated texture under my fingertips of a horizontal stack of densely packed cards. The ones for the core texts that everyone read, like Erikson's, were dirty with fingerprints, their top edges worn soft from frequent handling. Research articles were listed in minuscule font in huge tomes I could barely lift off the shelf.

When I recall my internet-free psychology studies, my mind jumps to other realms of life then, to how different things were. The World Wide Web was scarcely a twinkle in Tim Berners-Lee's eye.[3] Crossing Europe alone as an emergent adult, I visited the American Express office in each major city to ask if I had any post from home. My parents would have loved daily confirmation that I was alive, but it wasn't practical. I was operating on 30 American dollars a day and didn't have spare change for the payphone. I wrote travellers cheques and consulted atlases. When I lost the address and name of someone I'd met on my travels, I knew I'd never speak to them again. I shot several rolls of film, which seemed like a lot. I think I might still have the negatives somewhere, in a shoebox in a cupboard.

When I describe these wonders of the ancient world to my teenaged daughter, she looks at me as if I'm an elderly alien from another planet, and she's not entirely wrong. Courtesy of the digital revolution, there's a world of difference between Generation X and Generation Alpha.* But although she views me as clueless about how things are today, about that she's not entirely right. Since my early days in a pre-digital learning environment, I have indeed kept

* See Appendix 1.

Erikson's twentieth-century ideas alive for my psychology students and psychotherapy clients, but I've also studied the intersection of psychology and twenty-first-century technology for two decades. My last major project, a book exploring what happens to our online personae after we die,[4] really got me thinking. I realised I wasn't just interested in how a person shapes their digital identity, which remains behind online when they go. I wanted to understand the other direction, too: how that online world and our use of digital devices shape identity.

In other words, I wanted to consider whether Erik Erikson's map of the human lifespan still applies today, when so much has changed in the space of a generation.* I had both personal and professional experiences as motivation. Personally, I'm acutely aware of being in Erikson's middle-adulthood stage, which he called Generativity vs Stagnation. Midlifers can either keep pursuing their ambitions and goals, or they can get stuck, and this normal crisis took on abnormal proportions for me during Covid-19, when all my work and most of my socialising shifted online. I wouldn't have generated this book if I hadn't been so keen to avoid the double stagnation threatened by turning fifty and pandemic lockdowns. At midlife, I find myself looking forwards and backwards, sandwiched between ageing parents and a teenager whose various phases of childhood I can still acutely recall.

Professionally, I'm struck by how much my clients' dilemmas and troubles now relate to technology. People feel addicted to and fatigued by their screens. Parents stress over their children growing up digital. Romantic partners snoop through one another's devices when they're feeling insecure. Employees come to see me upset about productivity monitoring and metrics at work, or worried

* See Appendix 2 for Erikson's eight-stage psychosocial model.

about artificial intelligence rendering their jobs obsolete. People have their identities and narratives upended through revelations unearthed via commercial genetic testing. The bereaved are traumatised by losing access to a loved one's digital footprint, or by what they find out when they've accessed a bit too much.

Thinking about myself, my friends and family, and the clients in my clinic, I've realised that technology is the mediator and middleman in nearly every relationship we have in modern life, interwoven in the bonds between parents and kids, teachers and pupils, romantic partners, employees and bosses, midlifers and their elderly parents. We're constantly making decisions about data and devices that affect not only us but other people – often our nearest and dearest. Ensnared by our digital habits, we can't seem to wriggle free, and we fall into thinking and talking in particular ways about the technology that surrounds us – especially developments that disrupt, rattle or even frighten us.

In my roles as a psychotherapist and cyberpsychologist, I'm rarely asked how technology is affecting us, or how we're using it. People use a different word: *impacting*. We speak of the digital world as an active, powerful, unstoppable force, and frame ourselves as the victims: passive, without choice or agency. 'What is this technology *doing to us*?' people ask. 'What is it doing to our relationships, kids, sleep, communication, self-image, attention, grief, privacy?'

I contend that we're not as helpless as we assume, that technology influences but does not determine us. When we believe that tech has a monolithic impact on us, that it has unilaterally shaped our experience and behaviours, we're bound to feel powerless. In today's world, we often feel we've lost control. But we consistently underestimate the power we still hold to shape the relationships we have with and through technology, and to intentionally adopt the kind of mindsets that will help us negotiate psychological and social challenges at every stage and to reclaim our lives.

In searching for a useful structure to help you do just that, I reconnected with my old friend Erik Erikson. The wisdom and insights of his life-stage model still speak to us today, and I'll introduce you to them more fully in each chapter. Still relevant too – perhaps more than ever – is his concept of ego identity: the synthesis of your different 'selves' into one coherent identity over time, generating an ongoing, solid sense of self to help you navigate a world that is constantly changing around you.[5] And that's the thing: so much *has* changed since Erikson's time that his ideas about identity development and thriving need a revamp for the digital age. That's why I've created a whole new model – psychological, social *and* technological – to sit alongside his.

In this book, using this new roadmap, I hope to provoke and empower you.* I plan to disrupt commonly held assumptions and fears about technology's impact, suggesting a different response to technology and its challenges at every age and stage: a response that connects you to your particular context and your own values. For each phase, from digital gestation to digital afterlife, I'll explore how and why you might be pulled into using and thinking about tech in particular ways, even when those ways aren't helping you. I'll try to awaken your curiosity and awareness about why you use your devices the way you do. I'll push you towards gaining greater clarity about how well those choices are working out. And, most importantly, I'll invite you to commit to aligning your digital habits with the things that really matter to you, and the kind of life and relationships you want.

---

* See Appendix 3 for Kasket's Rebooted Techno-Psycho-Social Lifespan Model, 2023.

Digital life contains all the possibilities – good and bad, miraculous and disastrous, healthy and unhealthy. What you won't get from this book is one-size-fits-all rules or guidance, because you are uniquely you, and tech is incredibly diverse and doesn't have predictable impacts on individual people or specific relationships. Instead, you'll gain a better understanding of your *own* experience of tech. You'll come away with the internal tools and strategies you need to make tech serve you better, and the knowledge to assess the personal and relational consequences of choices you're making more honestly and realistically.

Fretting endlessly about how technology might be hurting you, or the ones you love, is a recipe for helplessness. The online threats and digital harms you read about in the news might make you anxious, but what if those stories are inaccurate or scaremongering, or simply don't apply to your circumstances? This book offers a fresh perspective that will prompt you to rethink your personal digital world, to take action, and to achieve a healthier, happier life in the process. Let's get started.

1

Digital Gestation

Springtime in Sawmill was dry that year. The previous summer's tall prairie grass, browned by the Arizona sun, rustled in a high wind. At the slightest encouragement, anything could turn into tinder, and southern Arizona locals know to exercise extra caution on such days. Dennis Dickey would probably never have started a campfire on a day like that. But that isn't what he and his friends were doing in the Coronado National Forest.

As with most special occasions in modern life, the moment is captured on video with the intent of sharing it on social media.[1] A rectangular target sits amid parched grass, against a backdrop of stubby, shrubby trees and distant hills. A shot rings out, and the target blows apart in a profusion of powder spewed skywards with the power of Tannerite, a highly flammable compound. The clouds of dust are baby blue: the sex of a yet-to-be-born child has been dramatically revealed.

Dickey can't celebrate the thought of a future son for long though. In milliseconds, the surrounding brush is awash in flame. By the end of the next day, 7,500 acres of land had been consumed by fire, and ultimately six times that acreage was blackened by raging flames that cost $9 million to extinguish.[2]

A state away, in southern California, blogger and mum-of-three Jenna Karvunidis wept as she watched the news. 'I was just

1

bawling when the forest burned down,' Jenna says. 'A cake isn't good enough. People are exploding things.'[3]

The piece Jenna had written years earlier about her 'gender-reveal party'[4] had swept the globe the way internet sensations do, but it had featured a far less pyrotechnic gimmick. In her article, she'd described cutting into an iced cake to reveal the colour of sponge inside: pink for a girl, blue for a boy. But followers of the trend had found ever more creative and bizarre – and occasionally fatal – ways not to settle for cake. An alligator snapping down on a watermelon, spewing arcs of blue goo.[5] An 'inadvertent' pipe bomb.[6] A novelty signal cannon.[7] A plane that pitched into the sea with its doomed cargo of two pilots and a sign reading, 'It's a Girl!'[8] One expectant father never made it to the party – the 'unspecified contraption' he was preparing detonated as he tinkered with it in the garage.[9]

The foetus, snuggled in the dark of the womb and preoccupied only with the business of physically growing, is innocent of all this. Even at this stage it is social, genetically programmed to be so – twins *in utero* reach out for one another more often than they direct movements towards the self, supporting the 'social wiring hypothesis'.[10] Outside the mother's body, the foetus is already socialising too, courtesy of its parent-created proxy, its digital avatar. As illustrated by the stories of drama and woe above, some of these are delivered via viral video.

'Surrounded by friends and family, you suddenly discover the sex of your child in this ritual, the very first identity marker,' says Tama Leaver, a Professor of Internet Studies at Curtin University in Australia.[11] '*This new person will be gendered in this manner, and identity therefore can begin!* It's a strange ritual.'

Strange, indeed, and stranger all the time. But the moment is ephemeral. The explosion fades away, and the cake will be eaten and forgotten. It's not the party itself but the publication of the ritual online, and the effects that this has, that will continue to be influential for identity formation throughout the course of the future human's life.

Erik Erikson and other life-cycle theorists of the pre-digital era never included the gestational phase in their models. Before the advent of the internet and social media, before millions of infants emerged from the womb to converge with digital twins that had been cultivated by family and friends for months, the time spent *in utero* didn't matter as much for identity. Now, however, there's rather a lot more going on in the prenatal period, and the digital milieu has created a novel context with new consequences.

For as long as we have had the means to determine the biological sex of a baby ahead of the birth, families and communities have engaged in prenatal gendering processes, painting nurseries particular colours, buying certain clothes, deciding on names that their society deems appropriate for a boy or a girl. Parents have spoken about their dreams for their as-yet-unborn child, and have made predictions about their talents, characteristics, interests and future profession. Maybe some of these parents even wrote these hopes down, in diaries or journals the child might see one day. What's so different now?

Digital networks have four designed-in features that offline social networks don't. Any information that's uploaded on them is persistent, replicable, scalable and searchable. In other words, the data we share last, can be copied, can quickly grow in reach and impact, and are easily found by those with individual, organisational and commercial interests. Professor Sonia Livingstone, a prominent scholar of children's rights in the digital age, says that this changes everything: 'People must contend with dynamics not usually encountered in daily life before . . . They include the imagined audience for online posts/performances, the collapse and collision of social contexts, the blurring of public and private spheres of activity, and . . . the ways in which messages spread within and across networks.'[12]

Performances. Parties have attendees and performances have audiences, and 'gender reveals' almost invariably have both. Events once confined to the family and its inner circle are widely consumed – unsurprising, as they were likely concocted and executed with dissemination in mind. A large and riotous crowd of adults and kids was present at one of the above-described reveals, gathering rather too close for comfort to the enormous, open-jawed reptile. Each person aiming their smartphone at this tableau – every visible adult present, except for the alligator wrangler – was almost certainly planning a far wider audience. You can still see the video everywhere online, including on a special YouTube channel: '2M Baby Gender Reveals: #1 Channel on showing the best gender reveal ideas. Videos Every Week!'[13] This first identity marker for a male child, future scion of Kliebert & Sons' Gator Tours,[14] took place in the new world of 'networked privacy'.[15]

Children have always entered the world carrying expectations upon their tiny shoulders, but predetermined impressions of them have never been so elaborated, so communal, so extensively networked and archived. Data about them have never been so available for consumption by close and distant, friendly or unfriendly audiences. The creation of this pre-established template may or may not eventually hinder an individual child in finding their own way, but an extensive digital public record has a weight that's hard to shake off. Authoring your own story is more difficult when the first chapter – and perhaps some of the later ones too – were written and made public before you even came to be.

But even more critical for future identity development, perhaps, is the early establishment of parental habits and practices. What we see others doing online powerfully affects what we share ourselves, and the publicising of prenatal identity markers and information on the internet has become so normalised as to be often done unthinkingly: biological sex, sonogram scans, relatives' names, hospital-visit details, planned names for the infant, due dates and data from pregnancy-tracking apps are all commonly posted online. If we're operating on automatic pilot, our human tendency to align with the behaviours of others in our social groups will drive our decisions, and early parenting practices will solidify into habit. Sharing data about a yet-to-be-born child, if done repeatedly, gives birth to new personal norms and forges parenting styles and philosophies that may persist for the longer term.

Obviously, the developmental tension of this new life stage is navigated not by the yet-to-emerge individual, but by their parents-to-be and those adults' wider networks and communities. Depending on which way parents lean during this new phase, either suspending prejudgements or making public predictions about identity, an anticipated child could be a blank screen or have a full-blown digital twin. The latter eventuality could provoke all manner of unintended consequences, leading to things that couldn't have happened in the era before social media. The story of Jenna Karvunidis, the woman who invented the 'gender reveal' party, is a tale of this kind.

Jenna had blogged virtually since blogging began, in the late 1990s. As a single woman, she wrote about dating. When she got married, though, she shut that blog down and started a new one. 'I don't like that digital footprint,' she says. 'I didn't need that stuff online.'[16] Still, when Jenna fell pregnant, she still had a large following.

She shared pregnancy-progression photos, images shot from the side to show her growing bump, and talked about ideas for the baby shower. When she hatched a novel plan – a party where she'd reveal her unborn baby's sex by cutting a cake – she blogged about it. The piece was a hit, picked up for a magazine that was in all the obstetricians' waiting rooms in the US at the time. 'It was my only time as a centrefold,' Jenna jokes. Pregnant women everywhere read about her gender-reveal cake and thought, *What a great idea!*

The phenomenon spread like virtual and then literal wildfire. After yet another scorched-earth tragedy decimating huge swathes of forest, a critic piped up on Twitter (now known as X). *Who the hell came up with this gender-reveal business?* One of Jenna's followers replied. *Oh hey! I've got your girl.* Suddenly journalists around the world wanted to talk to Jenna about the monster fad she'd created. Many of them came to the interview with assumptions about what she'd be like, and her response to their questions surprised them. 'I Started the "Gender Reveal Party" Trend. And I Regret It,'[17] read the headline in the *Guardian* in the UK. 'Woman Who Popularized Gender-Reveal Parties Says her Views on Gender Have Changed,'[18] said the leader for the National Public Radio story in the US.

When Jenna cut into her first child's cake, the crumb was pink. In the image accompanying many of the news stories, that child stands in the centre of a family portrait, nattily dressed in a blue suit and sporting short-cropped hair. 'PLOT TWIST,' Karvunidis wrote. 'The world's first gender-reveal party baby is a girl who wears suits!'[19]

The media stories portrayed her eldest child's non-binary iden-tity expressions as the reason behind her mother's change of heart, but Jenna tells me that her stance wasn't actually new. She was never trying to make a statement, she says; she was doing some-thing she thought was fun at a party, at a time when gender wasn't the hotly contested topic it is now. 'It was never supposed to be about gender in the first place,' she tells me. 'You can't know that at birth. You can only know what their body parts are.'

Jenna's tale embodies so much about how the internet works. She shared a personal story in a public network at a particular context, at a certain time. Although she didn't mind followers, she wasn't aiming to go viral; she was just writing about something she'd done that week. She omitted her actual views on gender from her article because she wasn't thinking about them and, anyway, the magazine wanted a lifestyle puff piece about parties. But people made assumptions, which is inevitable. Wherever a story is lacking, an army of creators and consumers on the internet will piece one together, even if it's wrong.

So it was that, years after her post, Jenna became an unwitting and unwilling poster child for some of the most contentious debates of the day: sex and gender, nature vs nurture, and what forces influence identity. She says she's grateful for the opportunity to set the record straight about her personal views, and to publicly advocate for what she believes. If she'd had a crystal ball, though, if ever she'd thought a gimmick she invented might contribute to a situation where it would eventually be harder for her own or any child to explore various aspects of identity for themselves, she probably wouldn't have done it.

Some might dismiss Jenna as a professional blogger and frame her information-sharing motives as profit driven. This wouldn't be the whole story – she's a law student at the time we speak and considers herself an activist.[20] In any case, though, most expectant parents aren't professional bloggers, and their sharing about their baby-to-be doesn't have global impact unless a disaster ensues.

Instead, any online activities pertaining to their unborn child exert powerful influence more locally, co-opting families and communities into co-authoring a set of predictions about who and what the child will become.

Martell Jackson* is a retired professional athlete, depicted in poster-sized black and white on the 'Home of Heroes' wall in my hometown, powering down the field like a steam train during an American National Football League game, the ball clutched in his hands.

His first two children were girls, both grown into young women by the time he and I connected on Facebook. The elder was super smart but not sporty, which Martell found hard. When his second daughter showed the prowess but not the passion, that was harder. He pushed her, trying to build up a fire in her, but it didn't work, and he had to accept that it wasn't her path. Martell's long-held dream of throwing a football in his backyard with a son was dwindling before he met his new partner, Jayda.

At the gender-reveal party for their first child together, a gigantic balloon hovers above their heads. To guard the secret until the critical moment, it's opaque and black. A pop, a flutter of blue confetti, and the room ascends into chaos. At the centre of the jubilation is Martell, who's punching the air with his fists, leaping ceiling high, and hugging his fiancée tightly enough to induce early delivery. Someone near the mic of the camera, perhaps the person holding the phone, is screaming, 'LET'S GO! LET'S GO! LET'S GO!' like a sports superfan cheering the starting line-up as it trots onto the field.

The video spread, with many people posting it on social media, including Martell. 'Check my reaction!' he says, adding a string of

---

* Names have been changed throughout this section.

cry-laughing emojis. 'This is how you respond to finding out it's a boy after 2 girls and 18 years.' Somewhere among the revellers are the two girls in question, but I can't spot them or their reactions to this joyfully unhinged, blue-tinted celebration.

Facebook posts from and to Martell in the coming months celebrate the anticipated male child's future sporting skills. Days after the announcement, someone sends a onesie and knitted hat with the emblem of Martell's alma mater, the place his football career began in earnest. Someone jokes about how mad he'll be if the sonographer is wrong. Someone else posts a cheeky photo of a skirted cheerleading outfit from the same university. *Until the child decides their gender identity, keep those options open.* The accompanying emojis suggest rolling-on-the-floor-laughing, not a pointed critique – he's winding Martell up, and the expectant dad responds in kind: an unimpressed-face GIF.

Each gift is systematically photographed and displayed, and onesies are standard, in various designs: the Nike swoosh logo, Martell's university insignia, the phrase 'Destined to be Drafted'. On Martell's own birthday, someone gives him a T-shirt emblazoned with BEAST and the image of a huge barbell, along with a miniature onesie that says *BEAST IN TRAINING.*

At points, Martell seems reflective about the culture of expectation he's helping create. He posts that his future son doesn't *have* to play sports but follows that by pointing out that he and Jayda both received top-tier sports scholarships, and that they're not the only high-flying athletes in the family. *DNA is real,* he says.

His friends and family concur. Yep, it's in the blood. That DNA is strong.

No doubt, I think. No doubt the DNA is strong. But in this nature-vs-nurture contest, with the baby's due date still some months away, the race is already looking pretty neck and neck.

When we speak, their first son is two and Jayda is pregnant with another boy, and most of Martell's social media posts are about how his sons will continue the family tradition of sporting glory.

Martell admits he and Jayda have had tough discussions about what will happen if the boys don't want to play. He hastens to reassure me, several times, that they'll love the boys no matter what. But there's a lot riding on it. For Martell, some of what's at stake is deeply personal. 'I had a decent sports career,' he says. 'But what a lot of people don't know is that I feel like I didn't reach the level I should have. I would like to see him eclipse anything I've done. I didn't make it. I want *him* to make it.'

The first image I ever saw of Martell's first son is on Facebook. They are in front of the wall of heroes in our hometown. Martell's chest swells with pride, and he is cradling his tiny son, careful to support his neck. Their matching T-shirt and onesie read *Team Jackson*. Martell's expression is triumphant, as if his son has made the team already, as though all the potential futures that exploded into possibility when that blue ticker tape floated down have already come to pass.

Martell isn't thinking about that in this photo. He's conscious only of this bond, this blood, this legacy, this love. He wants to show the world the two of them, together. 'There is no such thing as an infant,' said one of the founding fathers of developmental psychology, D. W. Winnicott.[21] 'If you set out to describe a baby, you will find you are describing a baby *and* someone. A baby cannot exist alone but is essentially part of a relationship.' For Martell, the expectations and the relationship are intertwined, and he has no shame in showing either. His favourite expression is, 'I SAID WHAT I SAID.'

Scrolling through Martell's and Jayda's Facebook feeds, I have my reservations about their posts, partly because of concerns about what will happen to all these data – more on that in later chapters. But I'm a digital-age parent too, someone who shared sonograms, and who posted an image on Instagram of my child's future name, colourful magnet letters I'd arranged on a smooth white surface. Looking at Martell's proud and joyful face, I understand his motivations. I have been there.

The concept of the 'blank slate' is as old as the seven hills that surround Rome. Few ancient residents of that city could afford to overuse expensive papyrus and parchment, so instead ancient Romans scribbled with a stylus on a reusable wax-covered wooden tablet, scraping the surface smooth again when a notation was no longer needed. In Latin, it's a *tabula rasa*. Later, Freud would use that term when he theorised childhood relationships as key in the development of your personality. Parent–child interactions, he said, form the deepest inscriptions of all on your personal wax tablet.

On the blank slates that were Martell's and Jayda's male children before they were born, their parents wrote their hopes – including the one that maybe, through these children, the hurts of their own pasts could heal and their dreams could be more fully realised. Online, the same children were blank slates until their parents etched their hopes there too. From that point, many others came to make their marks, to record their speculations. Marks perhaps also too deep ever to wipe away: long-lasting, reproducible, scalable, searchable digital marks.

Tama Leaver, who also blogged personal information about the first of his four children before he researched and reflected more deeply, shakes his head ruefully at the story of Martell. 'That kid, when they fail to get onto the high school football team, will be sitting in the shed, watching that video from fourteen years ago, terrified of telling their parents what happened,' he says.

One of Tama's research interests is sonogram-sharing – according to the results of a 2017 survey, more than a quarter of expectant parents share ultrasound images on social media.[22] After all, in today's pics-or-it-didn't-happen culture, if you're not sharing your sonogram on Instagram or converting it into a souvenir, are you even pregnant? One news round-up of ultrasound tchotchkes being sold online listed cakes, cupcakes, shower invitations, original prints, dog tags and lifelike dolls based on the scan image you submit to the website.[23] If you haven't got a good-enough image from the National Health Service (NHS), 'souvenir scanners' have risen in popularity on the high street precisely because 'some want imaging souvenirs, such as DVDs or keyrings'.[24]

When he and his wife were pregnant with their first child, Tama reckons they were among the last people in Australia to be offered a VHS cassette tape of an ultrasound. For the middle two kids, the clinic handed over USB flash drives, perhaps branded with the clinic's logo and website, but he can't remember for sure. By the fourth, they were provided with a bespoke social media platform and strategy. At their suggestion that hard copies might suffice, the clinic staff seemed flummoxed.

'They were like, oh, you don't want to do that! We'll send you everything,' Tama says. 'They SMSed the images, saying, *share with your friends!*' With one click, the images could be sent on to others. He describes the changes from first to fourth child as 'architectural', a word that conjures the profound shifts in the fabric and framing of our lives.

Sonogram-sharing began as soon as social media emerged, and

sonogram 'pimping' followed soon after. Writing about 'maternal devices',[25] gender studies scholar Sophia Johnson found a literal example: a 'Pimp My Ultrasound' app enabling you to add colourful baseball cap, cigar and speech bubbles to Victor's scan – or a tiara and tutu for Victoria, of course – and to share the results with your friends. Apps like this are marketed as a lark, a bit of fun, and of course they're that too.

But at the same time, something more significant is happening here. By customising the sonogram on the app, parents are again shaping a future baby into a particular kind of someone, setting expectations about identity that might or might not pan out in the future. On one hand, this encourages bonding by 'playing' with the ultrasound image, imagining it to have a personality, making it feel less abstract and more real. When the completed image is shared, others can join the game too. Everyone gathers around to reinforce one another's vision of what the baby will like and *be* like, as Martell and his friends did on Facebook.

It's Johnson who gave me the name for the resulting social being, a hybrid between the hidden, carbon-based life form in the womb and the visible, 'pimped-up' being that exists in the digital sphere. *Cyborg foetus*.[26]

Sharing anatomical photos of the *outside* of your body might result in your being banned from social media but sharing the *inside* gets past the censors. You might find it either fascinating or off-putting when people share scans of whatever sort, and some might consider it poor taste or too much information to display the interior of one's uterus. But it seems as if the consequences of such a decision would be levelled primarily at the excited parents-to-be – the occasional criticism, muting or defriending. But Tama describes how his sonogram-sharing back in the day was, in retrospect, 'a terrible idea'.

Why might it be a 'terrible idea' for the child? Where's the harm? At the prenatal stage, issues like facial recognition or safety considerations seem irrelevant. To the untrained eye, once you've seen one blobby, monochrome, vaguely baby-shaped sonogram, you've seen them all.

'Let's say you share the twelve-week sonogram on Instagram,' Tama says. 'You pull out the phone and take a picture of the screen at the clinic. Just a bog-standard sonogram has the mother's name, scan date, estimated date of birth, location of the clinic. All social media platforms process images by firstly saying, is there text in here? Can we pull this out? If it's in image form, it's [easy] to pull that information back out as text. Secondly, it'll look for metadata. Where was this taken? When? What other information can we extract from that?'

This is consequential, Tama says, because the moment Facebook or Instagram detects another person, even a person not quite yet born, they start a profile. Such platforms don't just profile the person who's decided to set up an account; they maintain profiles of anyone *connected* to them – child A and child B of This Named Person, whose dataset could easily be reconnected to their identity when they reach maturity. 'It starts to . . . build the idea of *so who is this, what do we know about them?*' Tama says. 'If that profile starts before they're born, [because of] the amount of information you'll already have on the system by the time they turn thirteen and sign up [themselves], that profile will be so much richer.'

One could argue that the data self will be born soon enough anyway, once the physical child *has* arrived, and that perhaps there's no downside in starting it a tad early. But, as Tama points out, any 'Coming Soon!' announcement creates an expectation in a community of close stakeholders and curious enquirers. If you don't spontaneously share when the actual event occurs, you'll be nudged to, increasing the possibility that you'll keep doing so. *Has the baby arrived? Post some pictures of the baby!* If parents who don't share information about their children online are unusual, expectant parents who start sharing prenatally and then go silent when the child is born are probably vanishingly rare.

'[Expecting a baby] is exactly the time you most want to share, and you're least likely to think through the consequences of that, years down the track,' Tama says. 'You're in the moment. And once you start sharing that stuff, it's hard to turn around and go, oh, now my beautiful baby is born, I'm not going to show you. You don't do that.'[27]

We're constantly nudged to share because the digital environment is designed that way. The so-called 'surveillance capitalists'[28] desperately want us to disclose information about both our identities and those of everyone we're connected to. Every scrap of data about you is hoovered up and commodified to the maximum extent possible, for power and profit. Google now knows you well enough to send you ads for things you're *contemplating* purchasing, without your having to search for anything online or say something out loud in a landscape bristling with ambiently listening devices. You might dislike this for the creepiness factor alone, but the manipulation often *works*. As mindful and in control of your choices as you attempt to be, you're still likely to be buying the baby gadgets, kitchen appliances and shoes you don't need. And you're an adult, with conceits of self-determination and aspirations to self-control.

If these impulse purchases are hurting anything, it's likely only you, your bank balance and, depending on what it is, the planet. But if sharing a sonogram today might unthinkingly establish a parental information-sharing habit that will eventually result in

15

a fully profiled teenager being served whole into the waiting jaws of surveillance capitalists sixteen years from now, perhaps that's something to ponder. You never imagine, when you're uploading an *in utero* image of your future baby, that this small action at such an early stage might render your child more effectively manipulated or exploited later. Any social media disclosure choices parents make about the pregnancy at that point naturally feel as though they concern only them.

'Relational maps of families and children are so valuable to companies because it's not just [about] what's happening *today*,' Tama says. 'The value of that pre-formatted young person, when they end up using one of these services, shouldn't be underestimated.'

One day, many people in a child's orbit will contribute to that little person's digital footprint. One day, the child will throw their own hat in the ring. At sonogram stage, though, parental control is near absolute, and the choice to share personal information about their child is largely theirs. Yet, caught up in the moment, unaware of the present and future reach and power of the audiences on the other side of their screens, over 25 per cent of parents share that image.

That statistic is old, by the way – ancient, in fact, on the digital-technologies timeline. Searching for and failing to find a more up-to-date number, I instead find myself directed to editable birth announcements on Etsy. Pink fabrics, pink baby shoes, blush-coloured flowers, baby clothes and a blush-coloured rose. 'We are excited to say, our sweet little lady is on the way.' Atop all these stereotypically girlish items is the space for the sonogram image. The listings say how many times the downloadable designs have sold lately, nudging the potential purchaser to pull the trigger. The images are clean, colourful, visually appealing and – importantly – square. Tailor-made for Instagram.

I wonder how many people now go straight from their scan to searching out a social media template like this. 'The challenge is getting through to people so they've front-loaded that thought process before they see that first ultrasound,' Tama says.

The thought process he means is deliberate rather than automatic, reflective rather than subconscious, aware of values and goals. Unfortunately, our digital environment actively undermines our awareness that we're engaging in habitual, learned behaviours – hence the phrase 'automatic pilot' – so we keep clicking, sharing, scrolling and liking. Habits don't have to be permanent, but to disrupt them you need to be sensitised to them, something that'll probably happen as you read this book. Before you can consider whether habits serve your goals and values, you have to notice them and become uncertain. Discomfort with how you've been habituated is a necessary prerequisite for change in a digital environment that's forever pushing you around.

But perhaps you've checked in with yourself, and you've decided that sharing your joy and your family news online is entirely consistent with what matters to you. If you want to feel deeply connected with your unborn child, and representing it online and sharing it with your community helps you do that, then such activities seem to meet your needs. When I was expecting, I shared plenty of words and images on social media about my future baby's sex, expected gender, due date and name. This was partly automatic, learned behaviour, but partly a conscious choice that I hoped would bind me more closely, at an important time, with far-distant members of my tribe. But I was a mother-to-be, and it didn't feel as if I was doing anything bad by sharing personally sensitive or identifiable data. It just felt like *care*.

Something else just feels like care in these times too: using technology to track and monitor your offspring from gestation forward. Prenatally, the focus of that monitoring can really focus only on growth, movement and health, and options for all that are limited between scans. But popular pregnancy-tracking apps gamify following the foetus's growth, instilling the idea good and early that part of being a good and loving parent in the modern age is keeping close tabs on what's going on. *This week your baby is the size of an avocado!*

Watching, storing, processing and sharing digital data about people has come to be known as *dataveillance*. Edward Snowden's revelations about dataveillance conducted by the US National Security Administration (NSA)[29] has given it a bad rap, associated it with capitalist and government bogeymen watching and controlling us for nefarious purposes, an invasion and exploitation of the less powerful by the more powerful. But a hefty percentage of the watching is conducted by parents on children, including the unborn. In that relationship, words redolent of Google, video doorbells and spies in the NSA don't quite fit. Deborah Lupton, an Australian sociologist who's an expert on the digital child, adds an adjective: *caring* dataveillance.[30]

While they're caring for the growing baby-to-be by tracking and surveilling, parents are also trying to care for themselves. They're scratching that itch we've all got now, the temptation to quantify all sorts of things we never monitored before, especially when those things have to do with one of our favourite subjects: our own bodies. The fetishisation of personal physiological data is the force that keeps you checking the primary-coloured movement circles on your FitBit or Apple Watch and pushes new parents to buy apps that enable them to size up their foetus against an array of increasingly large fruits and vegetables.

Pregnancy tracking might be fun, satisfying and reassuring, but it's more than that. When shared with others on Facebook or Instagram or other platforms, it gives mothers, in particular, a certain social capital. We've always expected parents to safeguard their kids and looked to women to be 'good mothers'. That role's always included hovering, watching, looking out for the safety and health of children both born and unborn. The digital age didn't invent that expectation, but modern technologies have kicked it up a notch. By sharing updates about your pregnancy with your online community, you're signalling something to others, something you might be feeling anxious about yourself. Maybe, in fact, you're using those technologies to assuage your niggling worries, your insecurities, your ignorance. You've never had a baby before, you're clueless, but it's okay – you're using all the apps. *This baby's in good hands. I'm a good mother. I'm a caring dad.*

Deborah Lupton sounds a cautionary note about caring dataveillance, a paradox: 'Any discussion of the ethics of caring,' she says, 'needs to acknowledge . . . these practices can both open and restrict freedoms for the watched subjects and those who engage in watching.'[31] Through caring dataveillance, some watchers might obtain freedom from social isolation or from worry. The watched, in this case the unborn, might be the ones who are restricted: in both the freedom to find their identity, and in their ability to protect it.

In a big-data world with rapidly advancing machine-learning and analytic capabilities, many uses of unborn children's information are yet to mature. When novel applications of that information are discovered and deployed, at a point too far over the horizon for us to see clearly, the consequences may race away from parents' original intentions as rapidly and uncontrollably as, in the wrong conditions, a spark bursts into flame.

Some types of data generated on foetuses, such as medical records and statistics that could be aggregated and analysed to identify and treat congenital problems, might reap huge benefits for society. Other kinds, including identifying information shared and digital avatars constructed pre-birth by expectant parents, are fed directly into an ecosystem that is forever seeking to exploit data for profit. The health insurance companies looking to spot and predict future health conditions and deny people coverage. The advertisers developing hyper-targeted marketing strategies, using everything they know about young people's weak spots to sell them what they supposedly need to be the person they want to be. The fraudsters, more skilled at impersonation every day with the aid of cutting-edge tech. Thanks to the information their parents and caregivers once shared on the internet, millions of today's children might find themselves fraud victims by the 2030s.[32]

Prenatally, and for a few years to come, a child won't be able to make its own data decisions, and cannot explore or shape its own identity. Parents and communities are their proxies, the only ones capable of asking themselves how their actions will shape the expectations that accompany that baby's emergence into the world. Parents can choose to hold the future personality in suspension, limiting how many identity markers and how much commentary they publish in the digital sphere, or they can make predictions that might become self-fulfilling prophecies; that might turn into conditions of worth that the child might or might not one day meet; or that compromise their offspring's future data security.

Actions tend to harden into practices when three things happen: when we do those actions a lot, when behaviours are triggered and encouraged by our environment, and when we fall into performing them as easily as breathing, without thinking about our goals or what's motivating us.[33]

Because the social media environment is tailormade for habit formation, so many of our digital parenting habits are formed prenatally: disclosing personal, intimate information about a new person who has no input or agency; tracking the foetus through the 'datafied baby' on your phone; including your future child in the information you regularly share within your networked public; and using information about that child in ways that gain you social capital, even in ways that constitute performance, such as a gender reveal. That social reward might not be consciously on your mind, but as with many habits, the subconscious is in the driver's seat, setting a course that so often can take you away from your goals and values – if you let it.

2

Infancy

The Owlet website radiates reassuring colours: light cool greens, whispers of beige.[1] Certain words are prominently displayed and repeated like mantras, balm to a new parent's ears. *Sleep. Rest. Peace. Easy.* A scrolling selection of quotes from relieved customers features at the bottom of the landing page, containing descriptions of the anxiety and exhaustion they once suffered, now so far behind them. Navigate to the 'Why Owlet' page, and you'll stumble into a gallery of cuteness: toothless smiles, chubby cheeks, bright eyes.[2] Accompanying each image is a story.

The rising action is subtle, conveyed through foreshadowy adverbials. He was a *seemingly* healthy baby. *At that point* her oxygen levels were fine. *Initially* he was breathing okay. Through these subtle linguistic intimations we understand, with a chill, that the tiny helpless baby is in mortal danger. If the parents are lured into a false sense of security now, if they make one wrong move at some imminent moment, all will be lost. A video running silently on loop on the home page captures the creeping psychological dread of this phase of the narrative. A mother opens the door into a dark bedroom. The hallway behind her is bright, but not enough to illuminate the corner where the child's cot sits. The mother's brow creases, and her eyes fill with terror. Is the room *too* quiet? Is the blanket moving?

The story's climax is the stuff of nightmares. Her oxygen levels dropped. His eyes were open, but he didn't respond. I put my hand on her chest, and it wasn't moving. He wasn't breathing. The parents shake their children, scream their names, yell to their partners to call 999, and rush pell-mell to the hospital.

But then the action falls. Each child in the photo gallery is wrested back from the brink of death. After the abject horror of that night, the babies grow up happy and healthy. The parents become calm, rested and productive, with all their past traumas healed and all present and future worries spared, because the resolution of the tale is always that the precious, vulnerable child never again goes to bed without their £289, blood-oxygen-monitoring Owlet Smart Sock.[3] While the app glows green, parents can relax. Across fifty testimonials on Owlet's UK website, the phrase 'peace of mind' appears thirty-two times.

Welcome to parenting through the interface.

Previous generations have somehow managed without it, so do parents need 'baby tech'? When you rarely venture more than a few metres away from your non-ambulatory child, how much value can an expensive bit of gadgetry, whether a new-fangled baby monitor or cloud-connected sock, really bring to your life? And how can digital tech have much influence on tiny people who can't track images flickering across a screen with their eyes, or grasp devices in their minuscule hands? From the baby's vantage point, infancy must be a pre-digital Garden of Eden, a time of innocence before that first Apple.

Erik Erikson said that the developmental task for the baby is navigating Trust vs Mistrust. The baby needs to trust its carers – and by extension the world – but the carers want something to rely on too. Before we had fancy monitoring technologies, harried parents turned to elders and more experienced peers for guidance. Then they started trusting distant experts: doctors and psychologists began telling us how to parent from the eighteenth century onwards.[4] The American paediatrician Dr Benjamin Spock published the first childcare advice bestseller just in time for the post-Second World War baby boom,[5] and eventually parenting advice became a multi-billion-pound industry of warring gurus.[6]

The opinions of these experts have varied about which infant-care methods best set babies up for life. In one corner, we have a camp advocating discipline and timetabling, arguing that strict sleeping and feeding schedules and letting babies 'cry it out' will ultimately render children more resilient, independent and able to soothe themselves.[7] In the opposite corner we have the camp encouraging proximity and close, supportive contact between parents and babies, known as sensitive/responsive parenting.[8]

The jury's not out on this: it's the latter parenting style that has been shown to be most beneficial for babies,[9] but it's not always easy to deliver. Round-the-clock sensitivity and responsiveness require a lot of time, presence and flexibility, and the social and economic world today is faster paced and more demanding than it was when some of the original proponents of responsive parenting were preaching their gospels.

In mid-century Britain, for example, paediatrician and psychoanalyst Dr Donald Winnicott regularly took to the radio to encourage 'ordinary devoted mothers' to parent instinctively and go easy on themselves for imperfection – 'good enough mothering', he said, was optimal for babies.[10] But employment outside the home was increasing, and mothers who tried to combine work and parenting faced a lot of stress and often harsh judgement.[11] By the liberated 1970s, when Penelope Leach picked up the torch as a responsive-parenting advocate, she got pushback. One newspaper retrospective of her career tagged her as 'legendary for making mothers feel guilty'.[12]

So, historically, even being a 'good enough', responsive parent while simultaneously maintaining your sanity sometimes feels like solving an impossible puzzle. But what if technology could enable you to have both?

Now that we have mod cons for infants, the utopian promise of today's baby tech is that it can make you the ultimate expert on your baby; create an environment that is exquisitely reactive to their every need, even if you're a parenting novice; and have an easier life for yourself into the bargain. With *babyveillance* – the portmanteau word for infant-monitoring technology – perhaps we don't have to settle for 'good enough' parenting. In an optimised twenty-first century, where there's a technological solution to interpret, antici- pate and respond to your child while making things simpler for you, why not do it?

Well, that depends on whether baby tech really facilitates the kind of care that matters most for those babies, and it's up for debate whether today's monitoring results in your infant being merely surveilled or truly seen. The new developmental challenge for wired-up babies, closely related to Erikson's Trust vs Mistrust, is Connection vs Isolation.

Baby tech claims to be capable of incredible things to make parents' lives easier and babies' lives better. Trawl the internet, and you'll find all sorts of innovations designed to assuage the worries of Gen Alpha's parents.

With infant wearables such as the Owlet smart sock, you can receive heart rate and oxygen-saturation reports on your phone, beamed through the Cloud. Smart cribs such as the SNOO and the Sense2Snooze[13] detect your baby's cries and respond with movements that mimic someone stroking their back or rocking the cradle. The Cubo AI Sleep Safety Bundle, which proclaims itself to be the first AI monitor, tracks microscopic movements, produces sleep analytics and alerts parents if the infant becomes stuck rolling over or gets its face covered in blankets.[14] The Nanit Pro combines calming white noise and lullabies with a climate sensor.

The Cry Translator smartphone app claims to discern, based on a five-second recording, if a baby is hungry, sleepy, uncomfortable, stressed or bored – the kind of insight I would have paid thousands for on a memorable transatlantic flight.[15] By understanding a cry's true meaning, you'll mount a more rapid and effective response, the website says. And that might not just matter for that moment, it continues – it might matter *for the rest of your child's life*. Your child will be more confident! More trusting of its caregivers! He or she will enjoy 'improved emotional development', linked to 'a lifelong increase in cognitive capacity'! While there's a gaggle of similar offerings on the App Store, Cry Translator is the only one I saw that classified itself as 'medical'.[16]

If a cry translator app reveals a child to be hungry, help is at hand. The description of a 'smart feeding bottle' from Annanta Baby, Inc. claims to facilitate closeness with your baby through WiFi and Bluetooth, and to 'reduce the transparency gap between parents and caregivers through seamless data transition', formerly known as talking to one another.[17] The aggregated wisdom about your little darling has everything you'd want and some things you didn't know you needed – a temperature monitor, 'IBM Watson analytics to track and monitor baby growth conditions in real time', and something called a 'poop library', which is either a source of diagnostic information or an opportunity to input outputs, if that's what moves you.

With the aid of these miraculous machines, parents can observe, interpret and analyse their pre-verbal infant in previously unthought-of ways.[18]

As a first-time parent, I suppose I developed the sixth sense mothers are meant to have, but my fears were never alleviated completely. How could I trust myself? I was a novice. Where parenting a baby was concerned, I was a babe in the woods. My own mother was thousands of miles away, I had no experienced elders handy to stand in for her, and I didn't know which baby expert to follow. If these high-tech, medicalised gadgets had materialised on my computer screen at 3 a.m., promising to relieve my suffering and give accurate insights into my child in real time, I would have been all in. Mysterious, otherwise-invisible vital signs beaming green-is-good reassurance to a glowing base station on my bedside table, and to an app on the phone that was always by my side? Yes, please!

When I first contacted Professor Victoria Nash of the Oxford Internet Institute (OII) to discuss baby-tech trends, we swapped notes on parenting our daughters, close age mates born around 2010. By the time our family didn't need its old-style baby monitor any more – if ever we did – Cloud-linked, Wi-Fi-enabled, smartphone-connected baby surveillance systems were in their ascendancy. No wonder I struggled to get rid of our monitor on the local sell 'n' swap Facebook page.

'After millennia, after generations of monitoring children while they slept by touching a forehead or listening to breathing, we suddenly need a smart sock or smart mattress or some other sort of device,' Vicky says. 'Where do the pressures come from? Why did this narrative emerge?'[19]

One source is obvious: the siren song of good old-fashioned advertising. Savvy marketers have long used babies to appeal to our basic instincts – neural imaging has confirmed that baby faces activate the emotional centres of the brain, eliciting our interest and attention, care and protectiveness.[20] Over the decades, infants have been used to convey safety, trustworthiness and family values in flogging everything from cigarettes to beer to car tyres.

Use those same babies to hawk infant-safety products to fledgling parents whose biological instincts to protect and serve their offspring are in overdrive, and you're in big business. Tama Leaver, the Australian Professor of Internet Studies who was quoted in the previous chapter, describes the new infant wearables as 'one of those fascinating cases of inventing a product and then convincing the market that it's an absolute necessity'.[21] It's what the advertising industry does best.

'The claims these companies make are very much presenting a narrative about being a better parent, a calmer parent, a more relaxed parent who sleeps,' Vicky says. 'They pick up on threads that parents really care about at these stressful moments: having young children, being sleep deprived.'

With my own child's early infancy many years in the rear-view mirror, the Owlet testimonials on the website still give me an apprehensive feeling in my chest. I know they're curated first to scare the life out of me, then to offer me relief: *Our product will remove your fear.* As nonsensical as it sounds, as the rested parent of a tween, with my eyes wide open and my critical faculties engaged, this kind of marketing can still work its magic on me.

But for most eager purchasers, affordability is likely to be an issue. In the US, Tama tells me, Owlet has offered an instalment plan to enable service personnel to buy it, but its price tag is not unusual. In *The Week*'s September 2022 consumer round-up of the best baby monitors available, your bill comes in between £166 and £369.[22] To shell out the amount usually associated with a major household appliance on a monitor, you'd have to be pretty concerned about your child's health, persuaded about the device's value or under the impression that 'good parents' spend as much as they can on babyveillance.

When you're feeling helpless or uncertain, it makes perfect sense to want to seek reassurance, especially if you don't trust your instincts. You consult Dr Google, look in on the baby, check with your partner (again) to see if you should be worried, record and then pore over the data on an app. Each time you get a hit of reassurance, you feel better.

But there's a problem. Reassurance-seeking 'works' through negative reinforcement – negative in the sense that something, like fear, has been taken away, and reinforcement in the sense that you're that much likelier to do the reassurance-seeking again. The stage is set for you to become addicted to an impossible quest for certainty – and certainty is the prize the babyveillance marketing dangles in front of us. Before you know it, you're monitoring and micromanaging every aspect of your baby's world with gadgets, perhaps to the detriment of your natural instincts.

As adults we're now accustomed to using tech to monitor our own daily lives. Before apps and devices were invented to monitor sleep, blood sugar, heart rate, oxygen saturation and exercise minutes, humanity managed to survive. But presented with the opportunity to transform one or more of these into graphically appealing charts, given the chance to learn about and gamify bodily functions, the benefits of the knowledge appear self-evident.

And yet, how does the monitoring make you feel? When you're winning the game, when the graphs go down or up, when the targets are met or the exercise circles closed, you're attending to your health. If the data don't look good, you're guilty or worried. But if the picture on the phone looks okay, that's the verification that *you're* okay.

We've put all our trust in this tech. My smart watch can't diagnose a heart attack in progress or alert me to an impending one, any more than the Owlet can predict or prevent Sudden Infant Death Syndrome (SIDS). But faced with an ambiguous sensation in my own body, I trust the technology to interpret it for me. When I bought my watch, I had no reason to suspect my heart was anything other than healthy, but I reasoned that if a new wearable technology could alert me to something I'd otherwise miss, that had to be good. Besides, there's little that fascinates and preoccupies us more than knowing about ourselves.

Except, perhaps, knowing about our kids, as evidenced by one piece of research that demonstrated the plunge in women's use of self-tracking and increase in infant-tracking apps once baby arrives on the scene.[23] Parents who have no specific concern about their newborn baby's health, but who see little downside in something that monitors their child's heart rate and oxygen-saturation levels, are as vulnerable to the promise of new tech as people with no cardiac issues who are nevertheless induced to upgrade to an ECG-capable watch. Monitoring such health information feels neutral at worst, lifesaving at best, a way to avoid those terrifying testimonials on the Owlet website. It's there *in case*, one of the innumerable mini-strategies we subconsciously deploy daily to stave off existential dread about ourselves and our tiny babies. The utopian promise of baby wearables is that each technological advance grants us greater freedom from mortal terror. Surely an enhanced sense of prediction and control makes you a better parent because you're less anxious. But does babyveillance really give you that?

We know that three-quarters of purchasers purchase the Owlet for 'peace of mind',[24] and we also know that the device doesn't always deliver on that – parents widely report that it sometimes has the opposite effect, especially when false alarms sound in the night.[25] What we haven't gathered as much data on is its impact on parenting practices. Vicky points me to the first research study on parents' actual experiences of Owlet 'in the wild'.[26] The study was small scale but in depth. The British mothers participating had no prior knowledge about the device, which wasn't advertised or commercially available in the UK at the time. When they first unboxed it, several mums felt let down that it didn't do *more*, such as measuring body temperature, something these new parents cared about during a project being conducted at the height of summer. Some would have preferred it to be functional all the time, like the mother who put it on her daughter for a car journey before realising it wouldn't work away from the home Wi-Fi network. But these parents, all of them mums, persevered for the two weeks to cast light on something that had never been assessed before. Would using a high-tech wearable monitor like this make any substantive difference to them – or their child?

At the end of the fortnight's experiment, the parents who participated had what the researchers conducting the study called an 'informed but casual attitude' about the sock, believing it hadn't done much to their parenting style. The researchers, however, had all the information from the study and a helicopter vantage point from which to analyse it – and they disagreed. What *they* observed was a surveillance device having an impact that the parents themselves didn't fully realise or notice.

First, for the parents, the sock didn't perform well on the anxiety-alleviation front. 'On the one hand, people liked the potential certainty,' Vicky says. 'The objectivity of this data monitor gave them extra reassurance. But in some ways – although these devices market themselves as providing reassurance for the parents – they might just generate new forms of anxiety and stress.'[27] Mothers who were anxious at the study's start – including those whose babies had some medical issues – were most drawn to the Owlet. But false alarms were frequent. When the app turned red, or the base station emitted an alarm in the night, maybe it didn't mean anything sinister. Perhaps the sock was put on incorrectly, or a kicky baby had thrown it off. But these events, or unfathomable fluctuations in the numbers, couldn't *always* be explained away, and such mysteries commanded more anxious attention.

Second, as the researchers put it, the device separated parents and infants in interesting ways. For one, just as we often become more obsessed with our health data than with health itself, using the sock diverted a significant proportion of parental attention from the actual baby to the datafied baby represented on the linked app. One mother described sitting with her partner monitoring numbers as their baby slumbered in another room. *Oh look, now it's 108. Oh, now it's 105.* The researchers called this 'fetishised tracking as a co-parenting activity'.[28] Another mum was so fascinated with her baby's vitals – normal, and always within the expected range – that she fell into checking them every five minutes. 'This is quite addictive,' one mum wrote in her research diary. Parents used the app to be more certain about their babies, doing experiments and testing hypotheses using the sock. Is Roy as light a sleeper as Lora thinks he is? Yep, readings confirm it. Could Sarah increase her own amount of sleep and work efficiency if she worked out how Lily's vital signs mapped onto her being ready for bed, or about to wake up?*

The Owlet app seemed to shape how people partnered with other people in caring for their babies. All the participants in the study had male partners, none of whom had downloaded the app themselves, but who frequently enquired about their babies' vital

---

* These are all example names, not actual names associated with the study.

signs with the mums, the main data gatekeepers for their children. If there was a 'failure in quantifying the baby' – a gap in the data stream, a puzzling or outlying bit of data, a warning that might or might not be a false positive – this became a conundrum for parents or people in the wider sphere to chew over together. Often the baby tech caused tensions and differences of opinion within couples over whether the gadget and its data were necessary, or beneficial, or accurate. Because the data were so sharable, mums forwarded bits and pieces of it to their wider network, setting a 'sharenting' trajectory early on. (We shall return to this in the following chapter.)

The researchers in the 'Quantified Baby' study showed how baby-tech data streams acted for the parents as both a reassuring and compelling virtual tether with their infants, shifting them towards deriving information about their child's status more from the data and less from actual contact. The chance to track, monitor and analyse data on the phone reduced the frequency with which the parents checked on the physical baby. Put another way, the digital baby on the phone competed with its physical counterpart for the parents' attention.

If that trend continues, as it very well could, we might need to be concerned. Expectant parents are already receiving strong messages that they 'should' use it, that it's as good as necessary – one mum described to me how her friends had been absolutely astonished and almost judgemental when she said she didn't want an Owlet, which they had been intending to buy for the new arrival having pooled their resources.* As Gen Alpha gives way to Gen Beta, due in 2025, baby tech is only becoming more sophisticated and artificially intelligent, encouraging our reliance on tech to help

---

* Anonymous, mother of infant of less than one year, September 2022.

us care for tiny humans. One day not long from now, nomophobia – anxiety about separation from our smartphones – might come to strike more fear into our hearts than separation from our children. And what will that mean for the babies?

In the cage is a tiny rhesus monkey. His brown fur sticks up from his head; his eyes are dark and wide in his adorably wrinkled face. Having been separated early from his biological mother to be raised in a laboratory, he is alone. Sharing his space, though, are two odd figures, both with plastic-and-metal robotic-looking heads. One figure is constructed of cold, hard wire mesh, but it has something the baby monkey needs to survive: it is equipped with two mechanical milk dispensers where a real mother monkey's teats might be. But the baby is clinging instead to the other figure, which provides no food, but which is swathed in warm, cosy terrycloth fabric. Sometimes, when he becomes too hungry to stand it any longer, he stretches across to access the wire mother's milk while hanging onto the terrycloth mother with his feet. If a startling noise sounds or a strange object is introduced into the cage, the little creature flees immediately to the terrycloth mother, gripping her soft but inanimate body with all his might.

When Harry Harlow first conducted his now-famous experiments in the 1950s, the prevailing view among psychologists was that food, not physical contact, bonded parent and child and ensured the infant's well-being. Many child-rearing experts believed that parents should actually limit their affection and presence, lest it spoil the child. But Harlow's work turned the tide, showing how critical physical closeness and comfort are for normal development, and how insufficient the mere meeting of basic needs. In some experiments, Harlow fed the baby monkeys and provided other conditions to maintain their health but gave them no company at all. With neither real nor dummy mothers of any material, some showed disturbed behaviour, circling their cages in distress and mutilating themselves. Some died. Others crouched in corners, removing their cloth diapers and cuddling with them.[29] In the decades of research since, we've learned far more about the specific ways that early physical contact matters. You can easily see that being touched, hugged and carried soothes babies, but less visible are what's happening inside and the lasting physical benefits that come from early contact. We now know that affectionate touch or its absence can cause 'epigenetic' changes – alter the way the baby's genes work. Infants who receive skin-to-skin touch in their earliest days have less reactivity to stress a year onwards.[30] Positive physical touch is associated with a healthier heart, gut and immune system later in life.[31] On the other hand, baby animals experimentally separated from their mothers, like Harlow's rhesus monkeys, tend to have inhibited growth despite being adequately fed.[32] Heart-wrenching studies of institutionalised, touch-starved Romanian orphans have demonstrated just how physically devastating being isolated from touch and attachment figures can be over the longer term.[33]

Healthy physical development isn't the only benefit of human touch for babies, though. Skin stroking promotes comfort, safety and trust. Although stroking from an object like a soft brush has some impact, it's not as effective as skin-to-skin connection in producing surges of the bonding hormone oxytocin – just as the wire mother was better than nothing but a poor stand-in for a flesh-and-blood monkey mum.[34] Being tickled, and the giggling and laughing that result, flood a baby's body with dopamine.[35] As infants come to associate the effects of feel-good and social hormones with close-contact physical interactions such as these, they learn to be human.

Following on from Harlow's experiments with monkeys, British psychologist John Bowlby[36] applied the findings to humans, and his ideas were so influential that they helped the public turn away from old Freudian-influenced and behaviourism-driven ideas about the primacy of food. Bowlby said that a consistent, close, reliable presence created a 'secure base', a healthy attachment between parent and child that would continue to influence the ability to bond with others throughout life. While it may not be a completely foolproof method – there's a complex interplay among genes, environment and attachment – if caregivers are sensitive and responsive, and if they have mutually enjoyable interactions with the baby, secure attachment is far more likely.[37]

If a baby has a secure base, as an older child they're likely to have less misbehaviour; more ability to regulate their emotions and soothe themselves in times of stress; a greater tendency to focus on the needs of others; more enjoyment and attunement with others in their social relationships; and more empathy. Babies with insufficient or inconsistent connections, or who are subjected to negative, painful touch, are more likely to grow up with mistrustful, self-protective instincts in response to conflict or difficulty; a tendency to either dominate or withdraw from others; and higher probability of psychopathology and behavioural problems.[38]

The researcher Mary Ainsworth identified that these issues clustered into categories when she watched how different children reacted on their caregiver's return after they'd been left in a room alone or with a stranger – these were the 'Strange Situation' experiments that produced the attachment categories we know now.[39] The securely attached children were pleased to see their caregiver come back and quickly re-established connection with them, but the less securely attached kids varied. Some were indifferent; some mixed clinginess with rejection; and some behaved erratically, froze or ran away. Ainsworth believed that how parents had responded to these children in their infancy predicted these behaviour patterns, and her Caregiver Sensitivity Hypothesis says that insensitive or inattentive caregivers may raise avoidantly attached children; inconsistent caregivers are likely to raise ambivalently attached kids; and abuse or neglect can lead to disorganised, fearful behaviour, such as the baby rhesus monkeys displayed when they had no kind of mother at all.

While the majority of children in recent generations have shown secure attachment patterns, parents relying more on baby tech and less on being physically present might change that. Ensconced in smart, sensitive, responsive nurseries, infants could become more indifferent to their human caregivers, perhaps producing a higher percentage of avoidantly or ambivalently attached children – and eventually adults. Down the line, there could even be a worst-case sci-fi scenario where parenting through the interface becomes tantamount to neglect, although this latter eventuality seems less likely. Loving parents who employ baby tech do so as an adjunct to their caring, rather than as a substitute for cuddles and physical presence. Technologically, we're a long way off from being able to delegate the care and handling of infants to robot minders, even if we wanted to. Harlow's studies focused on total disconnection from the real monkey mothers, and the Romanian orphan example, too, is extreme.

Still, parents might be sliding down a slippery slope when they're tempted by an advert for a smart bassinet with Cry Technology, which promises to respond to a baby's cries with 'front-to-back motion, soothing sounds and varying speed and vibration levels'.[40] If we start relying too much on technology to carry out the sensing and responding for us, as machinery gets exponentially better at doing both, we might gradually come to see it as a like-for-like substitute for parental touch and eye contact. As the infant grows and their vision and coordination improves, this latter factor will assume greater importance – especially when caregiver–infant eye contact is already constantly under threat.

Erikson considered eye contact a key pillar of building trust in infants,[41] and it's hugely important for babies' socialisation with their fellow humans. The sensitive/responsive parenting approach involves physical contact, reflecting babies' facial expressions, and mimicking their sounds and words, but also a lot of staring into one another's eyes. Infants' deep needs and desires for that kind of connection are demonstrated in a classic psychology study called the 'Still Face Experiment'.[42]

In a video of the Still Face Experiment, a lively, alert baby is sitting in a highchair.[43] The mother kneels before the baby, playing with her, holding her hands, looking straight into her face eye to eye, and copying her sounds and facial movements. Dr Edward Tronick, the Harvard psychologist who's narrating, explains that the mother and baby are in synchrony, 'coordinating their emotions and intentions in the world'. The baby is engaged and happy.

Then, something changes. The mother, instructed by the researchers not to respond to the baby for two minutes, stops everything. She doesn't leave the room or move physically away, but she assumes a flat, neutral expression, as though she is depressed, or zoned out. The baby immediately becomes distressed, doing everything in her power to regain the mother's attention and interaction. She points, babbles, shrieks, waves her arms, reaches out, contorts her body in the highchair and finally begins to sob, heartbroken.

Her mother finally snaps back to attention, the baby recovers, and all is well again. Dr Tronick explains that when reparation is possible – when the baby can successfully get the parent's attention again, then things can be okay.[44] Ideally, these moments are fleeting, part of being the 'good enough' parent Winnicott assured us it was fine to be. But what about situations where parents are unavailable in this way for a longer period, through depression or illness? And what if they're not really down at all – just distracted by their smartphones, forever looking elsewhere, present with their infants but with nothing like the eye-to-eye interactivity level of eras past?

The research thus far isn't looking good. The expression we tend to assume when staring at our smartphones is strikingly similar to the one that babies react to so negatively in the still-face experiment.[45] Unsurprisingly, parental smartphone usage is being shown to increase both infants' bids for attention, and worsen their distress when that attention is difficult to obtain.[46]

The harder it is to secure and retain the parent's gaze and response, the more negative the impact may be. 'The results can be very tragic,' explains Dr Richard Cohen, in his commentary on a still-face experiment conducted with babies and dads. '[The baby] can have trouble trusting people, they can have trouble relating to people, and they can have trouble being calm enough so that they can explore the world and take part in the world. So, we know that those initial relationships, that initial responsiveness and interaction between the father and the baby, are keys to the baby's success as a child and as an adult.'[47]

As the baby matures cognitively and is more able to process and interpret what it hears and sees, its observations become as important as touch or eye contact. Developmental psychology pioneers, including the psychologist Jean Piaget and educator Maria Montessori, argued that babies have far more cognitive capability than we give them credit for. Now, neuroscience is confirming that babies are tiny scientists, observing the environment and people around them, taking mental notes, learning and making connections.[48] Each time an experience or phenomenon is repeated, an infant's learning is reinforced and strengthened.

Imagine an eight-month-old whose parents have the resources to wire their home with fibreoptic cables, the salaries to splurge on fancy devices, the inclination to comb through product reviews. At half a year of life, that baby will on some level understand they are constantly watched, and that the mechanical eyes, lenses and lights scattered about the environment are integral to this enterprise. The baby sees and reaches out for the pretty bars or circles of colour on the glowing rectangles its parents are so preoccupied with, not yet realising that the face on the app is its own, that these shapes and hues represent the functions of its own body.

Perhaps the baby is starting to realise more abstract things too, things it would take many years to be able to put into words: that in this world, safety is king. Absolute, perfect safety, created and maintained through streams of data, achieved through what all that technology gives us: knowledge, or at least the illusion of it. Through observation and osmosis, the baby is learning that it is normal and expected for humans to relate and respond to one another through these objects, and that sometimes, because of competition from them, it might be difficult to extract a response from their parents at all.

Tristan Harris of the Centre for Humane Technology has said we're all subjects in the greatest experiment that's ever been conducted on humanity. A lot of that massive experiment has involved technologies that hijack our body's reward systems to keep us chasing dopamine hits in anticipation of the feedback or reassurances we might receive from them – merely glimpsing our phone can produce dopamine.[49] Perhaps smartphones and babyveillance tech are a kind of perfect storm for parents, combining the reassurance-seeking loop with the dopamine-reward loop. Meanwhile, while parents are pleasurably obsessing over and feeling reassuringly connected to their cyborg baby, the physical baby might not be feeling the same level of benefit.

We worry a lot about the impact of tech on older kids, who are old enough to use gadgets themselves. But as smartphones become ever savvier at capturing and holding parents' gaze, as those parents increasingly trust tech to detect and respond to infants' needs, we simply can't lose sight of the overwhelming evidence that physical touch and eye contact are essential for young humans. Both strongly affect development of the neurobiological structures that will support physical, social and moral functioning over the longer haul. And by the time infants can watch and interpret adult behaviour, they'll be absorbing lasting messages – accurate or not – about just how important they are to the people they love.

Erikson's Trust vs Mistrust is as relevant as ever, from both the parents' and the babies' experiences and perspectives. The more parents trust the tech involved in interface parenting, the more they rely on tracking the physical baby's vital signs from a distance via the cyborg baby on their phones, the more the senses that parents once relied on won't get used. The instincts that might develop from a different kind of knowing – a slower, messier, five-senses, incomplete knowing – could atrophy. Even if tech could optimise everything, perfection has never been required in parenting, and in fact might not serve infants best. If the foundations of adequate love, connection and physical touch are present, 'good enough' mothering or fathering isn't deficient – it's optimal for healthy development.[50] In the parent game, aspirations to flawlessness often backfire – in this as much else, as Voltaire said, the perfect is the enemy of the good.

In a doomed and unnecessary bid for perfection, we'd be doing both parents and babies a huge disservice if we allowed the shininess of newfangled baby tech to essentially return us to the era when it was thought food, shelter and warmth were all that infants required, back before Harlow did his monkey experiments, Bowlby formed his attachment theory and Spock revolutionised the baby-advice industry with his notions of sensitive/responsive parenting. We cannot let smart devices gradually become Gen Beta's version of wire mothers – the result could be an isolated, socially impaired, physically unhealthy generation. We know too much now to allow ourselves to sleepwalk into that way of thinking again. The physical contact and interaction that make both parents and children happier, that are so critical for infant development, are simply too important to forget or ignore.

Modern technology is doing much to help babies by safeguarding and enhancing the lives of infants with medical conditions. One day it might prevent tragic phenomena like SIDS, currently beyond the capacity of gadgets like the Owlet sock. In later infancy and childhood, video calling can help form strong social bonds between a young child and far-distant relatives, to the joy and benefit of all parties. But in the modern tension between Connection vs Isolation, it's likely that the more parents use tech to try to achieve the former with their very young infants, the more those babies will experience the latter. Tech is an increasingly powerful force in determining whether a baby is merely surveilled or truly seen, during a formative period when being seen, and touched, makes all the difference.

Luckily, though, parents and carers still have the freedom to decide how much tech they buy, and how they use it – if they can just retain the presence of mind to make the kinds of wise choices that serve true mutual connection.

3

Early Childhood

'What if you were watched every moment of your life?' the trailer asks.[1]

In the town where the movie is set, 5,000 cameras track every movement of one man, the unwitting star of the world's most popular television programme. He has no idea that he was adopted at birth by the TV production company that now films, edits and broadcasts his life for the entertainment of an audience. His friends, lovers and colleagues are paid actors. His wife's habit of enthusing in glowing detail about the ingredients of their meals is all about paid product placement.

One day, the jig is up: the main character narrowly misses being struck by a piece of lighting equipment plummeting from a clear blue sky. Squinting upwards, he spots a rip in the universe. As he watches, the black slash in the atmosphere is hastily closed by unseen hands, employees of the invisible studio beyond the bubble.

As it dawns on him that his entire existence has been curated for the benefit of shadowy watchers, the show's protagonist questions his very identity. Does he possess a personality of his own, or only what has been shaped in or pulled from him by his bizarre circumstances? Whatever power he thought he had to direct his own life is now exposed as a lie.

As a piece of fanciful fiction, *The Truman Show* posed some thorny existential questions, but its fundamental premises were still implausible on its 1998 release.[2] 'It couldn't happen,' said the director, Peter Weir.[3]

In the ensuing decades, though, we've come rather closer to its being able to happen. An emergent genre in the 1990s, 'reality TV' took off in the first decades of the twenty-first century. In parallel, both capturing and sharing our own real lives with the world became exponentially easier. YouTube appeared on the scene in 2005, and 'vlogging' – sharing one's life via video – exploded as small, easy-interface hardware such as Flip Video was launched. *Simple to shoot, simple to share*, ran Flip's advertising tagline. The first iPhone was launched in 2007, with image quality, editing tools and other advanced features – weapons in the smartphone wars ever since. In 2022, Korean filmmaker Park Chan-wook shot a film in collaboration with Apple, *Life Is But a Dream*,[4] using only an iPhone 13 Pro.

Internet-connected cameras in every hand all the time have massive implications for our intimate relationships and society. In 2022, tech journalist Amelia Tait interviewed the first children to have been blogged and vlogged on social media from birth, exploring what it is like for them as they enter their teenage years and find their voices.[5] When it comes to their parents having published edited versions of their children's lives to date, those voices are ambivalent. Tait calls this subset of the Alpha Generation 'Truman Babies'.

Many videos posted by the parents of Tait's interviewees had been staged and were stilted. Apparently spontaneous moments were often extracted with cajoling or bribery, more time-consuming versions of *smile for the camera, honey!* The videos were constructions, but unlike Truman Burbank, the Truman Babies were usually aware of and in on the process. The capturing and sharing of *their* lives surely were neither as exploitative nor as harmful as it proved to be in *The Truman Show*. Right?

*

To understand the potential impacts of a *Truman Show*-esque childhood on an emergent psyche, it's worth delving more deeply into the developmental world of the very young human. Not for nothing do we call these early stages 'formative': during these years the foundations of how we see ourselves, others and the world are laid. Erikson divided the period between infancy and school into two stages.

First, toddlers. The word evokes the eager, determined energy of the post-infancy years, the halting bids for an independence greater than such a young human can manage. According to Erikson, toddlers are locked in a showdown between Autonomy vs Shame and Doubt.[6] No longer entirely physically dependent on their care-givers, they're finding their own way, developing critical thinking, making their own decisions, learning from their mistakes and responding to their bodies' needs.

Watching their children stumble forward under their own steam for the first time, some mums and dads find their *own* loss of control scary and difficult and react accordingly, turning overly protective or restrictive. Others find it easier to provide a secure, accepting base for kids to return to while also encouraging and rewarding their independent exploration.[7]

In toddlerhood, erstwhile baby scientists become even more efficient little sponges, soaking up information at a crazy rate and drawing their conclusions, including conclusions about themselves. A toddler builds a sense of self from the raw materials of reactions and responses from others. Kids integrate the behaviour and feedback of adults into the foundations of a personality, including 'conditions of worth': internalised rules and assumptions about what behaviours are good or bad, and whether you're acceptable or unacceptable.[8]

Through early-childhood experiences your 'early maladaptive schemas' (EMS) began to form – themes and patterns that will

47

enduringly affect you and your relationships, manifesting throughout your life through emotions, thoughts, memories and physical sensations. EMSs develop when core emotional needs are frustrated or go unmet, including secure attachment to other people; the freedom to express emotions; freedom and autonomy; and a sense of identity.

You might not be able to pin down the origins of your own EMSs and schemas, because they lie beyond the reaches of earliest memory, which struggles to access material before the age of two and a half.[9] But the vulnerable spots that trip you up today might very well have been initially sparked in those earliest years – not just by major incidents and events but also by subtle and innumerable micro-interactions; verbal and nonverbal messages; fleeting reactions and responses from the people in your life.

My adult psychotherapy clients are forced to make assumptions about the conditions of worth that were imposed on them in childhood. Working backwards from how they feel and think now, they construct stories about how their early maladaptive schemas likely came to be. They can't rewind the tape to check how things really played out.

But relative to previous generations, Alphas whose parents were active on social media have had an extraordinary amount of detail recorded about their early years. Technologies don't (yet) exist that let us peer directly into other people's brains, but Gen Alphas scrolling through their parents' old Facebook or Instagram feeds have more insight into their minds and hearts than I will ever have about my own folks when they were young parents. A Gen Alpha can bear retrospective witness to their parents' daily shifts in mood and attitude; the ways they talked about parenting and about their kids; and their implicitly and explicitly expressed hopes, expectations and judgements.

In some ways, Gen Alpha *can* rewind the tape. In a succession of screen-captured moments, they can observe the development of their own early schemas and conditions of worth, growing and unfurling like a plant on a time-lapse video.

Then there's the phase after toddlerhood, from three to five, for which commonly agreed-on words don't seem to exist in English. The only word that springs to mind references what is yet to come: *pre-schooler*. You might have recollections of this time in your life because the parts of your brain that support narrative and memory are better developed at this point. Erikson called this the Initiative vs Guilt stage.[10] Children become more independent, assert their wills, decide what's worth their time and attention, and push back against people imposing ideas on them. Conditions of worth and feelings of unworthiness are strengthened if carers are dismissive and undermining. Parents who support and create space for a child's emergent interests, though, can help the young person's sense of purpose bloom relatively unconstrained.

These relationship dynamics are standard issue in early childhood. As a parent, it's hard not to let your own hopes, fears and needs encroach on your kid – it's always been that way. Our core emotional needs are called 'core' for a reason. But when we take this familiar psychological recipe for early childhood and stir in a new ingredient – social media – what happens?

While they might find their way around many features of a smartphone or tablet, the youngest of children don't generally use social media in its current iterations, don't create their own profiles. Instead, they're represented by self-appointed proxies. Proud, doting, well-meaning parents have always displayed their children's pictures, crowed about their achievements and woven their own assumptions, expectations and projections into the narratives they create for and about their children. But – just as happens before birth – social media have redoubled parents' power, influence and impact in children's lives, sometimes for better, and often for worse.

To sharent, or not to sharent: that is the question.

In the compound-word world of tech neologisms, *sharenting* is the act of disclosing information about one's children on social media, a practice so common on the wired-up side of the digital divide that the *New York Times* once published a memorable piece entitled, 'If You Didn't "Sharent," Did You Even Parent?'[11] The accompanying video portrayed children of various ages engaged in conversations with their befuddled, somewhat defensive sharenting parents, most of whom were confused and resistant about the idea that there might be any sort of problem with the practice.

Parents aren't the only ones sharing information about kids on the internet. If a school or health care provider posts information about kids online, parents have every right to protest. But the waters become murkier when the sharers are in one's own family, and when there is a mix of non-sharenting and sharenting stakeholders. In 2020, a sharenting grandmother in the Netherlands hit the news over the photographs of her beloved grandchildren she'd posted on Facebook and Pinterest. The person requesting removal was her own daughter, the children's mother. When Oma steadfastly refused to cease and desist of her free will, the daughter took her to court – and won.[12]

What is it like for children to have their information captured and shared throughout their formative years, before they can understand these technologies, or have the agency to make decisions about their digital participation? How do sharenting practices interact with a child's core emotional needs, and affect the parent–child relationship? Largely because of sharenting, this time of life has acquired a new layer: Agency vs Powerlessness.

In January 2010, Michael Arrington was sitting onstage in San Francisco. The co-founder of TechCrunch, an online magazine about high tech and start-up companies, was interviewing founders at the 'Crunchies', an awards ceremony lauding innovators in the field. Arrington wore a suit, but the person joining him – to enthusiastic applause and whoops from the crowd – was clad in jeans and a zip-up grey hoodie.[13] With its galloping growth and astonishing revenue, this young man's company had won the 'Best Overall Start-up' award at the Crunchies two years running. The preceding twelve months had brought 50 million more users to his platform, which had recently updated its settings to make all social media posts public by default.[14] If many of those users weren't bothered by that change, privacy campaigners were certainly annoyed.

'You've always pushed the envelope on privacy,' Mr Arrington remarked, alluding to this controversy. 'Where is privacy on the web going?'

'Well, it is interesting looking back, right?' the young man said, waving his hand airily. 'When we got started . . . a lot of people asked . . . why would I put any information on the internet at all?' He went on to describe how those same people had acclimatised rapidly to sharing more and different types of information with wider audiences, including personal information in public spheres.

'That social norm is just something that's evolved over time,' he said. 'And we view it as our role in the system to constantly be . . . innovating and updating what our system is to reflect what the current social norms are.'

The man in the grey hoodie was, of course, Mark Zuckerberg, then and now the CEO of Facebook. Casually proposing that public was the new private, he argued his company was simply giving the world more of what it wanted by changing the default settings on Facebook. The revelation that privacy was no longer a social norm should have been a bombshell, deserving further discussion, but Arrington moved on to other things, asking whether Zuckerberg was planning to continue being 'aggressive' in acquiring other companies. It's a question to which we now know the answer. Two years later, Facebook would buy Instagram.

The claim that people didn't need, want or expect privacy was bold, and if true would help Zuckerberg reap a great fortune. His platform's business model, after all, rested entirely on users sharing as much information as widely as possible, about themselves and other people. Over the decade to follow, Zuckerberg pivoted this way and that in his public statements about the importance of privacy. His company was not *responding* to new norms of privacy and disclosure. Instead, Zuckerberg and his fellow entrepreneurs were *creating* those norms, with a lot of success, prompting countless debates over whether we are still in control of our choices and able to chart our own destinies, or whether technology has hijacked our will and stolen our freedom.

I'm motivated to understand both the programmed and social pressures to share information about children because of my own sharenting history. While it could be considered ironic, in this chapter I'll share my own stories in situations where I happen to be an expert by experience as well as by training. Parents need both compassion towards themselves and empathy towards their children to navigate a technological landscape that skilfully and subtly pushes people to share. The fact that I was compelled to do so, despite all that I know to be important in early childhood, is testament to that fact.

As Arrington and Zuckerberg discussed the death of privacy, I was far from Silicon Valley. I was sitting in a freshly painted nursery in London, cradling my infant daughter in my arms. Like so many others in my new-parent cohort, the class of 2010, I had announced her imminent arrival via a sonogram image on social media. My daughter's data self was born on Facebook, months before she physically arrived.

By the time she was ten, about a third of the world's population was reportedly using Facebook at least every month.[15] In January of 2020, Zuckerberg's company was preparing its press release for Data Privacy Day, and I was sitting over my laptop attempting to purge every post relating to my child on the two flagship brands now under the parent company of Meta: Facebook and Instagram.[16]

A decade's worth of social media proved surprisingly tricky to eliminate; Facebook was offering no bulk-delete option. At first, Instagram seemed easier. I scanned the tiles for her face, her distinctive hair. Having deleted every image of her I could spot, I did what I thought was a final check. But there she was again. Was something wrong with my eyes, or was Instagram playing cat and mouse with me, concealing and revealing secret caches of images? Another round of deletions, another review, and there she was still, in photos I could swear weren't there a moment ago.

I wanted to stop but couldn't. She kept looking in on me, to see how it was progressing.

I was engaged in this task partly because researching and writing a book about what happens to our data when we die had rattled me into thinking more seriously about digital life. But the tangible countdown to Deletion Day began in earnest about a year earlier, at lunch with my daughter.

She asked for lemonade. 'Maybe. I'd like to ask you something first,' I said. She cocked her head warily, wondering what today's exchange rate for soft drinks would be.

'I'd like to record our conversation,' I said. 'If that's okay. It's part of a project.'*

'You're not posting it, are you?' she said immediately.

Sometimes the mere stimulus of my drawing my phone from my bag would elicit this response. Her reflexive reaction to my question was the reason I wanted to talk in the first place.

'I read a story in the news about a teenager,' I said, recognising in my tone the forced casualness parents deploy in the hope that their kids will open up to them. 'She was talking about her parents posting photos and other things from her life on Facebook, and how she felt about it.'

I didn't want to prejudice her response by telling her more: that the eighteen-year-old Austrian woman had been so discomfited by the hundreds of photos her parents had posted – 'They knew no shame or limits,' she was quoted as saying – that she sued them in 2016, claiming they had violated her right to a personal life.[17] Or that in France the law allows adults to sue their parents for compensation over images posted in their minority. Even without these salacious details, my daughter was intrigued. She pushed her colouring page to the side.

---

* The following dialogue is drawn from the recorded conversation between myself and my daughter, which I have her permission to use within this book.

'I was wondering if you'd tell me your opinions and thoughts,' I continued. 'A lot of parents share stuff on social media before their kids are old enough to know anything about it. Tell me what you think.'

'I didn't like it when you were doing the funny conversations on Facebook,' she replied.

I was floored. She meant the mother–daughter conversations on various topics, dialogues I'd faithfully transcribed and shared with friends and family over the years. I'd titled them like Michel de Montaigne's Renaissance essays.[18] *On love. On nature. On memory.* People had adored them, given the thumbs up, asked for more. I loved them too, and since I'd printed them out in a physical book the previous Christmas, she'd seemed to enjoy them immensely, paging through them in bed at night. She hadn't liked them?

I held my tongue, noticing the effort it took not to defend myself. Having been given the floor for once, she told me all of it. All the times her trust had been betrayed, the times she asked me not to share and I did it anyway, the times she'd been surprised or angered by learning I'd posted without her knowledge. All the occasions strangers seemed to know everything about her, greeting her like an old friend. All the times I'd deprived her of the ability to decide her own boundaries. All the times she'd felt exposed.

I wasn't a professional blogger or the matriarch of an Instagram family. I wasn't Gwyneth Paltrow or Pink, celebrities with huge numbers of followers on social media, both of whom had recently been in the news for sharing information about their kids online without the consent of those kids.[19] Paltrow's daughter had called her out in the comments of a specific post.[20] I was a person in the modern world with a kid, not significantly different from any other parent who has social media accounts. Sure, my passions for photography and writing – and perhaps the 5,000 miles that lay between me and my hometown – might have prompted me to use social media differently from some, certainly to use it more. But my privacy settings were stringent and my motivations pure: share the love, keep in touch, warm hearts. Her take was different.

My daughter's memory for certain incidents was extraordinary, stretching over almost the whole of her life. She cited events from before the age of three, things she shouldn't have been capable of remembering at the toddler stage of cognitive development. She didn't sound angry, merely matter of fact. *Resigned*.

'Honey,' I said, 'I am so sorry. What would you like me to do?'

Her eyes lit up. Here was her mother suddenly offering her a choice, a choice that perhaps she should have always had. She seized this new power with both hands.

'I want you to take it all down, and I don't want you to post about me any more,' she said.

Something in me convulsed and recoiled. A dozen argu-ments came unbidden into my mind, reasons to deny her based on convenience, efficiency, memory and joy. Reasons to do with connections with faraway people we cared about. If I reassured her about my privacy settings, would that make it okay? If I reminded her how restricted my list of friends was, could I preserve my treasure hoard? I'd never kept a baby book. Instead, I'd built this lovingly curated online repository of beautiful photos and charm-ing conversations. It had been so easy, automatic, simple to record simultaneously moments for myself and her father – and for her, surely for her! – and share them with grandparents, aunts and uncles. Memories like that are sacred, I thought. You can't just destroy them.

But the question had been asked and answered. What choice did I have? My daughter's heels kicked rhythmically against the bottom of the banquette, punctuating the silence.

'When is this going to end?' she said.

The moment hung between us. She meant, of course, the recording. She'd allowed me to take a voice memo of her dropping the bombshell that the first decade of our relationship was coloured by my having taken such a liberal hand with her information, and now she was ready to move on.

'Now,' I said. 'It ends now.'

I'd stored the photos elsewhere, in physical format or on USB. I'd even downloaded my entire Facebook archive and had posts and images printed up in books, the size of each volume growing exponentially for each year I'd spent on social media. Still, with each deletion a pang of worry hit my chest. I'd been utterly conditioned to think of digital back-ups as necessary for the safe preservation of memorabilia and important documents. What if a house fire burned my albums and melted my hardware? Trawling through years of memories online, I was recalling things I would have forgotten entirely had I not captured them on Facebook and Instagram, and my stomach kept flipping. *What if I forget again, lose these memories forever?*

Maybe this discomfort has been designed in too, I thought. Maybe they're making it both procedurally and psychologically hard to erase my and my daughter's data. They *want* me to give up on my data-deletion project. Were these barriers to removal sneaky, resistance-is-futile tactics designed by Silicon Valley developers, people far cleverer than me?

Later that month, a piece that was probably prepared at the same time I was pursuing my deletion mission was released on Facebook Newsroom.[21] Data Privacy Day wouldn't exist if it weren't for social media. The annual awareness-raising event started in

2007, a few months after Facebook was unleashed on the public.[22] In his 2020 statement, Zuckerberg acknowledged Facebook had 'a lot of work to do' on its avowed intentions to build stronger privacy protections for the site's users. He reassured people that the 'Privacy Checkup' tool had been updated and the company was now being more transparent about information it collects about its users, both on and off the platform. The short article had a long headline: 'Starting the Decade by Giving You More Control Over Your Privacy.'[23]

Giving. You cannot give something that you don't have, that is not in your power to gift. If a social media platform can hand me control over my privacy, it must possess that control already. And, of course, considerations of my *own* privacy weren't the primary reason I was deleting posts that day, under the watchful eye of a ten-year-old who had gained the maturity and summoned the courage to assert herself. And I wondered what combination of internal needs and external influences had compelled me to create this avalanche of data in the first place.

The forces that drive parents to sharent are complex and interactive, but let's look at the recipe: a cocktail mixed of one part individual psychology, one part behavioural psychology and one part social psychology, flavoured with a dash of neuropsychology, poured into a glass of CHI (in this case, computer–human interaction). The individual psychology component involves one of the core emotional needs common to all humans: feeling secure and safe within your relationships; belonging; enjoying the approval and response of your tribe.

Being a first-time parent far from my family and place of origin, I was hungry for all of that. My friends and family might not have been able to routinely spend time with my daughter, which was painful for all concerned, but these pains were made so much better because I made her as vivid as possible for them via social media. The isolation was reduced, and the connection was strengthened.

That conduit of connection was as much a lifeline for many of them as it was for me. No wonder they rewarded everything I shared, and behavioural psychology dictates that actions that are rewarded are likely to be repeated. Content about my child received more likes and comments from my community than anything else, especially when it consisted of a fantastic photo paired with the transcription of a funny dialogue between us. Praise of my writing or photography supercharged the power of these reinforcements for other reasons: they validated valued parts of my identity as a writer and photographer, and I'd been frightened my identity would be lost or irrevocably changed through motherhood. Through curating the posts the way I did, I was ticking both boxes: I was a mother, yes, but my community let me know my previous identity was still there.

Human behaviour is influenced by both individual characteristics and social situations, and though we underestimate its power, it's the situation that's the stronger force. Especially when we're feeling uncertain, in a novel or ambiguous situation where we don't know what to do or how to act, we look to others to work out what to do – this is a phenomenon known as *social proof*.[24] If everyone's doing a particular thing, if the behaviour is conventional and expected, we do it. We tribal animals fall into step, taking our cue from others. Thus entrained, we pace ourselves to the prevailing direction and speed of traffic – including internet traffic.[25] And when we're in step with our tribes, it feels *good*.

Because influencing our behaviour is key to their products, technology companies bake all this psychological knowledge into the devices, platforms and services they design. Many of the founders and developers in Silicon Valley attended Stanford University, home of Professor B. J. Fogg's Behavior Design Lab, previously known as the Persuasive Technology Lab.[26] As much as we'd like to resist the incessant nudges from our technology, some argue that it's becoming close to impossible. The best minds in the world, as Google-employee-turned-philosopher James Williams has written, are focused on harvesting your attention, keeping you online for as long as possible, and getting you to surrender the maximum amount of monetisable information.[27]

But another psychological concept is relevant here. In the 1960s, a famous series of experiments was conducted at the University of Pennsylvania.[28] Illustrations of this study in textbooks always involve worried-looking dogs in cages, the floors of which were constructed of wire mesh. The dogs were concerned because these floors could deliver painful shocks. Any animal would work hard to evade such an aversive stimulus, and, indeed, the dogs tried, at first. In some conditions of the experiment, the dogs could escape the shock floor by leaping a barrier or pressing a lever. In the cruellest and most revealing condition, however, there was no way out. Try as it might, the dog could do nothing to get away from the shocks.

After a time, having learned the situation was inescapable, the dogs in this condition lay down on the floor. When the pain hit, they whined but did not move. They'd absorbed the belief there was nothing they could do, so when researchers tempted them with treats, threatened them with more punishment or demonstrated ways to escape, the dogs wouldn't budge. The lead researcher, Martin Seligman, called this resignation 'learned helplessness'. Learned-helplessness behaviour happens in humans, too. If you believe there is nothing you can do to change the situation, if you believe someone else is at the controls and there's no way out, why go through the effort of trying?

In the sharenting conversation with my daughter, I asked why she hadn't called me out on my sharenting over the years, hadn't driven it home to me just how much it bothered her. Why hadn't she told me?

She shrugged wearily.

'I didn't think you would stop,' she said.

As a sharenting parent, my relationship with technology wasn't working for my kid, and it hurt the relationship between us. But I understood why. Perhaps as you have been, I too was hypnotised and reinforced into data-sharing behaviours, in sway to the business models of big tech. There is absolutely no judgement here. But I knew I had to stop. As a psychologist, *my* business model rests on change being possible, and on finding the leverage to enable people to choose the change, to nourish what's important to them as individuals – which might include growing in power and combatting certain social and technological influences at work in their lives. For me, the leverage proved to be my willingness to acknowledge that my sharenting had interfered with some of my daughter's core emotional needs.[29]

Core emotional needs aren't just foundational for children. Every stage of life requires them to be met or attended to, for they form the foundation of all good relationships, helping us to live happier and more grounded lives at home and work, more confident in ourselves. Among the most critical emotional requirements is autonomy – a theme that, as we've seen, looms large in early childhood, when children are becoming their own little people, and parents are struggling to let it happen.

Whether you think parents and family should have ultimate authority over what's best for a child depends a bit on your culture as reflected in the laws of the land. In the UK, parents are largely considered the gatekeepers of their minor children's information, and the parents' right to freedom of expression often trumps children's rights to privacy. Although a child could use the 'Misuse of Private Information' (MOPI) argument to claim their privacy rights have been violated, success is far from guaranteed, and 'misuse' is hard to prove in a culture of widespread sharenting of personal information.[30]

Ironically, France – the same country where adults can sue their parents for material shared on social media when they were children – is well known for a parenting style where the grown-ups call the shots: *C'est moi qui décide* ('It is I who decides').[31] In the United States, the 'parental immunity doctrine' prevents kids from suing their parents in the interest of family tranquillity, meaning that kids will be unable to one day take their parents to court for childhood bathtub photos on Instagram.[32] There are exceptions to this immunity – if a parent has engaged in 'wilful and wanton' crimes or misconducts against their child, that child can seek redress. In sharenting, the current or future harm is rarely understood, much less done wilfully and wantonly. But questions of autonomy, and the consideration of whether people see their children as their 'Mini-Mes' or their own people, came to my mind when I noticed the consistent refrain in the *New York Times*'s 'If You Didn't "Sharent," Did You Even Parent?' video.[33] *But it's NICE to share my life, and you're PART of my life.*

Conscious of respecting their children's wishes, some modern parents don't sharent at all; or set arbitrary age thresholds for them to start consulting kids before they post; or at least back down and demur when their kids proactively or reactively object. Others negotiate or plead, offering incentives or extracting agreements to be able to post photos at a particular frequency or on specific occasions: once a month, at the start or end of the school year, on family holidays or occasions. And yet others ask their very young kids for permission.

I asked my own daughter about this, during the pub-lunch privacy summit. 'Would you have preferred me to ask you?' I said. 'Would that have been better?'

My daughter's tone was patient, if a little patronising. She used the example of a younger friend, whom I'll call Adrienne, famous for her willingness to let it all hang out. 'If Adrienne's mother said, *Hey, Adrienne, can I take a picture of your bum in the bath and post it on Facebook?*, she'd stick her bum at the camera and wiggle it and say, *Sure!*' my daughter said. 'But Adrienne's crazy. And she's a *child*.'*

She was right. Consent is meaningless if the person giving it has no capacity to understand the consequences of that consent. In the case of social media, parents are often in the same boat as the kids, for only true knowledge and understanding grant the full capacity to consent, and adults struggle to grasp the full picture too. Many people focus on a few obvious drawbacks to posting young kids' images on social media, remaining unaware of the subtler, less visible problems. Even when we have an inkling of abstract future consequences, in-the-moment concrete impulses often win the day. But the ways in which a child's present and future can be affected by sharenting are legion.

First, as previously noted, there are concerns that sharenting could spell financial woe for the next generation. Children's data are a goldmine for marketers, and facial recognition technologies will help the marketers of the future recognise and target them so expertly that it might render them helpless to resist what's being

---

* All material quoted from recorded interview between myself and my daughter, used with permission.

sold. Fraudsters will have a field day too – it's estimated that, by 2030, sharenting will have resulted in over 7 million cases of identity impersonation leading to financial fraud.[34] Thanks to today's data-sharing practices, the criminals of the future might find it far easier to take out lines of credit under the names of newly fledged adults, or to drain their bank accounts and digital wallets. Having experienced both the siren call of super-targeted online marketing and the inconvenience and cost of identity theft, I'd be concerned about unwittingly increasing the chances of making anyone else more vulnerable to such crimes, much less my own child.

Dignity and safety, too, are real concerns. The former Australian Children's E-Safety Commissioner, Alastair MacGibbon, was already reporting in 2015 that up to half of the images found on paedophilic image-sharing sites had originated on friends' and family's social media.[35]

'People have a kind of conceit that they have a lot of control if they are stringent about their privacy settings,' says Belinda Winder, a Professor of Forensic Psychology at Nottingham Trent University who specialises in sexual offending.[36] 'They tell themselves, *I know who my friends are. I only have people I know within this circle. I know what happens or doesn't happen when I put pictures of my kids up online.* But people forget that there's a whole household of connected individuals that are connected to each of those people on their friends list, or that there may be wrinkles that they're unaware of that enable those images to travel more widely.'

Many of those images, Belinda hastens to add, aren't sexual or questionable, but innocent photos that wouldn't violate any social media community guidelines at all. Elsewhere online, or on the dark web, such images can be placed into sexually exploitative contexts. If that makes for some uncomfortable contemplation, she goes further. Estimated rates of childhood sexual abuse of roughly one in twenty for children in the UK make for some uneasy maths.[37] Who's doing all this offending?

'Something like 80 per cent of sex offences are committed by someone with no previous criminal records,' Belinda says. 'They're not a *monster* monster. Ninety-nine per cent of them could be a brilliant person, and then 1 per cent of them you don't see, a part they're keeping hidden away in a dark, dark place that they are ashamed and embarrassed about as well.'

We'd like to assume that the people *we* choose to connect with on social media are nice, that everyone in *our* family and friends lists are safe, and that friends of friends must be reasonably okay too – hence that privacy option. Unfortunately, it's likely your online and offline circles contain sexual offenders.

Corporate exploitation of children's data, although not of the same stripe as sexual abuse, can also be consequential for children's lives. In *The Age of Surveillance Capitalism*, Shoshana Zuboff explained how the data we generate about ourselves and other people are fed into increasingly sophisticated predictive tools to anticipate our current and future choices.[38] The surveillance capitalists of the world sell the resultant, highly valuable 'prediction products' to marketers hungry for information about both current consumers and future ones – including children. She terms the big business of horse-trading information about future consumption the 'behavioural futures markets'.[39]

Those markets are big business, and with children's photos and other information so ample on parental accounts, the temptation for a company to monetise such data is strong. For example, Facebook's 'Scrapbook' function organises an individual child's images into a dedicated album on the parent's account.[40] Nice for the parent, perhaps, but a feature that's suspiciously convenient for the company offering it.

Privacy regulations try to protect kids from companies that care little about their individualism, their humanity and their long-term prospects. Parents *are* invested in these and yet don't fully understand how they can compromise these interests through the actions they take: not just sharenting, but through buying their kids data-gathering smart toys featuring facial recognition and AI.[41] And by creating vivid portrayals of their children for a wide online audience, parents might also be impinging on a core emotional and psychological need for this and every stage: the freedom to write one's own story.

Sharented kids have something in common with celebrities: *parasocial* relationships.[42] People think they know you, but you know you don't know them.

You're famous! said one group of my friends, encountering my child for the first time in her memory. She looked flummoxed. *What are you listening to these days?* they said eagerly, expecting her to list some classic rock stars. My post about her David Bowie-themed birthday party, bittersweet when it proved to have been held on the day of his death, had been extraordinarily popular in my circle. I'd been all too happy to present my muso daughter to the world. But her tastes had moved on, and she named something modern and poppy.

This response did not receive positive reinforcement. My friends' faces fell, conveying to the child that she had delivered the wrong answer, that there was something wrong with her current preferences or ways of being. Worried that they'd think I'd faked it all, or at least exaggerated her interest, I felt flustered and became pushy too. *Well, you still like lots of kinds of music, don't you?* I said.

She hadn't met the conditions of worth, conditions made present and powerful by my taking something that was once her, once hers, and sharing it in a way that fixed her supposed preferences and others' expectations of her. My actions limited her feeling of flexibility, her ability to change and develop without self-consciousness or guilt. *But aren't you more of a Bowie girl?* everyone said, referencing her identity rather than her taste.

I cringe at the memory. On the way home, she seemed troubled, even irritated.

Am I famous? she asked. *Why do they think they know me?*

While I'm concerned about the behavioural futures market, as a parent and psychologist I'm equally concerned about what I suppose I could call the social futures market – data published and traded among members of a community that can affect a child's social and developmental trajectory. Alphas are the first generation to be sharented from birth, so we have yet to see the impact of millions of children growing up surrounded by the assumptions and expectations of people they've never met.

Anecdotally, though, I know this: in sharing information about my own child, I cluttered up her field of possibility to a depth and breadth I couldn't have managed before the internet. In the wider social community that encompassed both me and her, I reduced the horizon of personal identity that she had yet to traverse and that she had every right to explore for herself. There was almost no room for her to be free. Instead of feeling autonomous, she experienced confusion and doubt. Instead of being able to take her own initiative in the conversation, she was instead forced to justify her current musical tastes and nudged into feeling guilty that she didn't fancy David Bowie so much any more.

C'est moi qui décide. I never even realised the decisions I was making. Such was the extent of my power. My child didn't explicitly say no to my sharing, but she did engage in what psychologists call *protest behaviours*, using indirect words and actions that pointed to her emotional discomfort.[43] Reading the dialogues I once posted on Facebook, I find the evidence of how I carried on regardless.

Are you writing this down? Putting it on everyone's iPhones?

The entire time I've been talking about the chicken farm, you've been writing. What are you doing?

Like an addict, I'd started to cover my tracks, transcribe under the table, lie about my activities. That's proof enough that on some levels I knew, and that she was no fool. At best, she learned that my interests superseded hers. At worst, she felt gaslighted. At a virtual open day for a secondary school when she was ten years old, she winced when the headmaster referred to their graduates as 'finished products'.

I am not a product, she said to me, betraying a sensitivity to being commodified that she did not acquire from any direct exposure to social media. She got it from me.

The internet has an amplifying effect, intensifying the impact of early-childhood issues that have long existed. Some of the consequences of that will come home to roost in the present, and some will emerge in the future. In that future, the child's own use of social media – whatever form it takes at that time – will give them more control. By that time, however, certain templates, patterns, schemas are already entrenched.

If a child feels as though their preferences about privacy and self-presentation matter less than the parents' needs, imagine *that* as a foundation stone in a developing personality.

When a child's story is written and framed for a wider audience of family, friends and strangers, even by a tremendously well-meaning and loving parental PR machine, it deprives the child of the chance to be their own storyteller. People thought they had my child pegged, compromising her freedom to unfold as she otherwise might have, to make her own choices. She wasn't the narrator of her story – she was the subject of mine.

So, for the digital Generation Alpha children of today and the Generation Beta children of tomorrow, Agency vs Powerlessness is huge. I don't like to think about the powerlessness my own child felt, although I think and hope I repaired much of the relational damage when I acceded to her request to delete my past posts and refrain from sharenting in the future. In opening the discussion, apologising for past behaviours, asking what she wanted and following through on her requests, I gave her back both voice and influence. I showed through word and action that I believed in her right to forge her own path, and promised to work harder to give her space to do that. I hope I admitted what I now consider to have been my mistakes, giving her context for my actions without using them as justification or excuses. In doing all this, I hope that I demonstrated humility and respect and that, looking back, she will remember me for that. I hope it's the kind of consideration she will go on to show others, both online and off.

Imagine the alternatives to sharing children's data automatically on social media. Imagine parents deliberately and mindfully finding other ways of meeting their own emotional and social needs. Imagine parents actively generating wider, creative options for sharing their own lives and their family lives with the people who matter, and mindfully resisting the technological pressures and nudges to share from a fuller understanding of the benefits and consequences. Not only to avoid the bad, or the potentially bad, but to move towards the good, or the potential for it.

Children are far likelier to thrive psychologically if they experience themselves, even in this age of sharenting and social media, as agentic narrators rather than powerless subjects. The more liberty children have to develop and explore their emergent selves, released from the watchful gaze of quite so many virtual others, the better. Yes, we could ask their permission to post, but we don't have to place such decisions on their insubstantial, unprepared shoulders, and we don't have to post ourselves. Instead, we could leave the question of how they wish to appear digitally to the world to them, a complex dilemma to be tackled further down the road.

4

Later Childhood

When I was growing up in American suburbia, people talked about *latchkey kids*, a phrase that wasn't created for Gen X but that definitively entered the lexicon when the Baby Boomers' children were young. I don't know whether my mother, whose work in those days was looking after us, pitied these children – perhaps some secret part of her envied those parents who went to the office – but I was jealous of the latchkeys. They could gobble treats from unattended cookie jars, ride their dirt bikes on streets they shouldn't, claim they'd done homework when they hadn't and watch TV in the afternoon without let or hindrance. They could do anything and go anywhere they liked, and if they were home and behaving themselves by 6 p.m., their parents would be none the wiser.

When my own daughter entered school, life was quite different.* We lived in an expensive urban area, both of us had to work, and paid carers were pricey. Every day, one of us would have to sacrifice income or an important meeting to pick her up at school. But around the time she hit ten, a note came home in her rucksack. *Parents who wish their children to leave school unaccompanied must sign this permission form.*

---

* Material used with permission.

Could our daughter really be a latchkey? This wasn't 1970s Indiana; this was London. I craved a security blanket, perhaps in the form of a security *system*, the kind unthought of in my day, except as portrayed on sci-fi programmes and futuristic Saturday-morning cartoons. I read one opinion article that described being left alone after school as 'formative to identity', which worried me, but perhaps that didn't have to be negative. Couldn't some autonomy be *good* for her developing sense of self? Mightn't she relish her independence, take pride in the trust we'd placed in her and develop strong resourcefulness skills? Plus, parenting through the interface at this age could be more of a two-way experience, I thought, not the unidirectional surveillance of babyhood. Knowing that her parents were aware of and caring about her presence in the home, able to communicate with us through the tech we installed, perhaps our daughter would feel contained and safe – not really alone at all.

In a few clicks a system that would enable a new way of life was researched, purchased and winging its way to us. A slimline video doorbell would capture her as her little fingers fumbled with the door keys, and a second device would verify she'd made it into the hallway. We could speak to one another, my disembodied voice issuing fuzzily from the doorbell or booming from the front room. At 3.20 p.m., sitting with a therapy client, I'd feel two successive vibrations from my smartwatch. Knowing she was home, I could be more physically and mentally present in my afternoon clinic than at any point since she'd been born.

When she reached eleven, the transition to a further-flung secondary school prompted a security review: the school would no longer be down the road, so the monitoring would have to travel with her. Guilt and uncertainty nagged at me as I considered the privacy implications, but I not only went ahead but made it non-negotiable. *You want to go to school and move about the neighbour-hood on your own? No problem . . . if we're tracking you.*

Every day, I pick up my phone multiple times to locate and imagine her. When her icon judders slowly down that street, she is dragging her feet on the way to the bus stop. When it lies stagnant in that corner of that building, she's in English class. When it zooms down that road, she is winging her way home on a red London double-decker bus. I assume she is safe, and a weight lifts from my shoulders.

From the first she wasn't a fan of it, the surveillance. Maybe the fruit was a leftover from lunch, or a handout from the after-school club. I saw it on the stored video clips later, as I reviewed the day's activity at my front door. Her little blue eye came level with the camera for a moment before she drew back and, slowly and methodically, ground a half-eaten banana into the lens.

When that first day of school arrives, everything changes. While family members may still be the top influencers, friends, classmates and teachers assume a far more prominent role in a kid's emerging self-concept. Novel, difficult things are suddenly expected of them: reading, doing sums, sitting still, paying attention in class and getting along with other kids. Gradually it dawns on a young person that being successful in the world outside their home means demonstrating certain traits and skills valued by society. Even a child who's received unstinting love and support from their families might feel rather less robust when they learn that approval and acceptance are often more conditional on the outside. Some rise to the challenge, emerging with feelings of competence and good self-esteem, while others struggle to stay afloat, doubting themselves and their abilities. Erikson captured this range of possibility when he named the ages between five and twelve the Industry vs Inferiority stage.[1]

Inside vs outside, home vs school, here vs there. In the twenty-first century, tech has blurred the lines dividing these formerly discrete realms. The demarcation between the 'inside' territory of family and the 'outside' space of school, peers, travel and outdoor play has broken down, because the new world of tracking and surveillance technologies creates a hyperconnected web of information linking once disparate spheres of a child's life and containing data that might have far-reaching implications.

As they wave to the video doorbell on the way to school, an AirTag dangling from their rucksack and their smartphone pinging with reminders from parents, as their school behaviour and performance is digitally tracked, they're absorbing messages about their goodness and trustworthiness; about the metrics that matter; and about how people see and judge them. Before children were so frequently electronically monitored, they experienced both the exhilaration and anxiety of exploring the world with a greater independence, getting into jams and having to figure things out on their own. Before nearly every error added another link onto a long digital chain, failures and mistakes were more easily forgotten.

But now, as digital-age children explore the big world outside their front doors, they are likely to be doing so as wards of a purportedly benevolent and protective surveillance state. A new tension now intertwines with Erikson's Industry vs Inferiority stage: Confidence vs Insecurity. Whether tracking and monitoring a school-age kid promotes one or the other isn't a simple question to answer, but one thing's for sure: as they use a panoply of devices and apps to watch and store children's data, parents and educators are shaping young people's beliefs about themselves, other people and the world in the process.

At school age, as it did in infancy, safety preoccupies carers' minds. Just as when their children were tiny, parents are once again acutely aware of the potential vulnerability of their offspring, more conscious of the limits of their own control, more tempted to manage the distance between their children and themselves proactively. And, as it did in babyhood, a greater level of surveillance starts to seem like not only a potential solution to these discomforts but like a normal, even expected practice for people who have the means to try to guarantee their children's safety. If you can afford the tech, why wouldn't you?

On closer examination, though, the parallels with infancy start to break down. For babies, dependence on their carers is a psychological and physical necessity for survival. For school-age kids, shedding that reliance is a major developmental milestone, but it's not always easy for parents to let go. Looking over my shoulder while trick-or-treating as a youngster, I spotted my dad, who'd been following at a discreet distance. I have no idea how often he or my mother physically tailed us, but Gen Alpha parents can do it from the comfort of home.

Still, they've got their work cut out for them, because the workload for the modern surveilling parent is massive compared with babyveillance. Heart rate, oxygen saturation and safe-sleeping positions have long ago been dropped as targets of concern; as school begins, worries about children's physical health become merely the tip of a very large iceberg. Below the waterline lies a more nuanced and wide-ranging territory, parents' anxieties ranging from unsuitable friends through bullying up to stranger abduction and death, but mums and dads seem to be up to the challenge. Coinciding with Data Privacy Day in 2022, the cybersecurity company Malwarebytes conducted a survey on parental child-tracking practices and discovered that 84 per cent of parents did it in one form or another.[2]

Age nine was the most common point at which children became the subjects of their carers' overt or covert tech-facilitated spying – the age my own child was when we installed the video doorbell.[3] The amount of monitoring increases as children age into the pre-teen and teenaged years, and just over a third of the nearly 900 parents who completed the survey reported that they had kept at least some of their watching secret.

Where are they? is the most common question parents pose to the tech about their kids, and apps can easily answer that without awkward conversations or surreptitious trailing. A 2019 survey found that 40 per cent of UK parents routinely used GPS to pinpoint where their child was, with 15 per cent admitting to doing so 'constantly'.[4] For the uninitiated, let me assure you that it's hardly difficult. Virtually all smartphones come with inbuilt child-tracking potential via functions such as Family Link on Android phones or Find My on Apple products.

If you want yet more capability, though, you can download a 'social networking' app actively used by over 30 million people that might seem less like socialising and more like stalking: Life360, which purports to 'protect and connect' every member of your family with the Family & Friend Locator installed on their phones.[5] Life360 has reaped such rich rewards from parents that it's now eyeing up new markets: besotted pet owners, for example, and vulnerable seniors.[6]

With Life360, you sit at the centre of the panopticon, able to see everywhere. When your child enters or exits the 'geofence zones' that you've drawn around your home, their school and anywhere else they visit regularly – their best mate's house, the local park – you receive a notification. Like the companies that produce and market babyveillance technologies, the people at Life360 clearly know that intimations of danger and promises to alleviate anxiety are the twin Achilles heels for parents. *Get ready to breathe easy*, the tagline on the website reads. *We offer peace of mind designed for modern life.*

The second most common question is probably *What are they doing?* What they're up to offline can be inferred from location data, of course, but of equal concern are kids' online activities and conversations. In 2020, 61 per cent of UK children aged five to fifteen had their very own tablet, 55 per cent a personal smartphone. Most kids watch YouTube and other video-sharing platforms (VSPs) from an early point – across all ages from five to fifteen, the percentage of kids who consume online videos never drops below the upper 90s – yesterday's television is today's YouTube. Nearly three-quarters of under-sixteens play online computer games, sometimes with the ability to chat with strangers online. Even with the most careful setting of controls, a child can still stumble across unsuitable websites.[7]

You might have the illusion that social media isn't a factor until the teen years, but these platforms aren't just the realm of the over-sixteens for whom they are intended: nearly half of UK kids aged eight to fifteen have accounts targeted at over-sixteens, and a third of them have an account meant for adults.[8] Although some kids act alone when they fake their ages to access these platforms, that's not always the case – parental facilitation is common. And while today's children might open an email with the same frequency adults receive letters via carrier pigeon, over 97 per cent of over-twelves use messaging apps[9] and would greet with hilarity the idea that – in Europe and the UK, at least – WhatsApp is only for over-sixteens.[10] If every parent and teacher followed the app's directions on how to report an underage account in order to have it closed, there'd be rioting in the playground at schools.

Such statistics make clear just how much there is to cover. Parents have no hope of overseeing all this manually for much the same reason that so few of us bother to read the T&Cs for the online platforms we use: who's got that kind of time? In addition, few of us have the appetite to negotiate oversight with kids who naturally grow more protective of their privacy at this age. So, many mums and dads find the answers in outsourced, automated, ping-if-there's-a-problem tech.

Qustodio and Bark are the most popular child-monitoring apps in the UK and US respectively, and the promise of ease that they offer is tantalising. At the time of writing, Bark hasn't been able to pass muster with UK and EU privacy legislation, and a lot of UK parents I've spoken with are eager for that to happen. Qustodio enables remote monitoring but isn't as automated and notification-driven as Bark, which 'runs quietly in the background' to detect any bogeyman or disturbing material that might cause or indicate harm to a child – predators, bullying or content intimating drug use or self-harm.[11] Bark is like a guard dog, always on duty. You don't have to be the bad guy, you don't have to spend time hovering, because Bark will keep you in the loop with as-needed notifications: *We've spotted potential issues.*

A video about Bark depicts mother and child sitting on a sofa, absorbed in their respective devices. In the old days, mum might have had no idea that the son was being bullied; maybe he would have been ashamed to say, not sure how to talk about it or he might have wanted to deal with it on his own, perhaps afraid that any action his parents might take would make the situation worse. But the family uses Bark, and a millisecond after a frown creases the boy's brow at the sight of his own screen, a text lights up his mother's. Turning to camera, mum explains that Bark enables her to respect her son's privacy by not having to read through every single message.

'[Bark] helps me and my son work together, trust each other, and build a deeper bond,' she says. 'Plus, their data is always protected and private, giving me further peace of mind . . . Bark brings us closer together.' She illustrates this closeness by putting down her phone and turning to her son. *Hey. Can we talk?* she says. *Sure,* he replies.[12]

Such a solution to keeping tabs on a young person's digital life feels like a parent's dream come true. It's not always easy to contend with a child's reticence, at a time when friends are often so much more important than parents for development of self-esteem. But could the same data that parents are accessing to keep their child safe be used by other forces to render them vulnerable?

Much has been said about who profits from the data we generate in the current system, and who might be disadvantaged or harmed by that situation.[13] There's no doubt that the potential victims include today's children, whose data are fuel to the furnaces powering the behavioural futures markets. Although Life360 pledged at the start of 2022 to phase out these arrangements, until then it had been one of the largest raw data sources for the location-data industry, supplying US military contractors and some of the biggest data brokers in the world. Often, the data they supplied was not even anonymised.[14]

That company is far from being alone in its business models and practices. We ought to be concerned about where such comprehensive and intimate information about our kids ultimately ends up, and whether it could eventually be de-identified in ways that could hurt them – socially, occupationally, educationally, financially or physically. Even when an app assures us – as Bark does – that the data are always protected and private, such a claim has more to do with what it says on the contract, rather than what happens in practice. Because there's such big money to be made from children's data, we should probably be cynical about assurances that companies are trying to make this asset *less* appealing to the greedy data brokers that want to buy it. The Common Sense Privacy Program, which evaluates smart tech used by kids, reports that Bark does not meet its recommendations for privacy and security.[15]

So much is already out there, though, that it's easy to feel discouraged, to succumb to the helplessness we've almost been trained to feel. When I spoke to Dr Gilad Rosner, the founder of the Internet of Things Privacy Forum, he chuckled when I said I hoped this book could help restore people's sense of control over their information.[16] He felt there was little point in worrying about the stable door after the horses have already bolted. 'Individual control over our personal information is basically an illusion now,' he said.[17]

Certainly, in the technological and regulatory ecosystem we're currently living in, the wider-scale issues we're facing seem to defeat any notion of personal agency, for us or our kids. Even the companies harvesting our information can't figure out where the horses have got to. Quizzed by a special investigator in the ongoing lawsuit over the Cambridge Analytica scandal, two veteran Facebook engineers were utterly stumped when asked precisely what information Facebook stores about us, and where it is. 'I don't believe there's a single person that exists who could answer that question,' said Eugene Zarashaw, one of those engineers. 'I don't know. It's a rather difficult conundrum.'[18]

If what Zarashaw has testified under oath is accurate, what hope for control or choice is there for the rest of us?

But there's power in thinking globally and acting locally. If we become sensitised to even a bit of the theory and research about what's important for healthy psychological development at the Industry vs Inferiority/Confidence vs Insecurity stage of children's lives, and if we can act on that, we'll find ourselves in a position of greater power.

*

Imagine a ten-year-old child, living in an inner city, the leafy sub-urbs, or a sparsely populated rural area. They often move alone between home and school, riding their bike between their own house and those of family and friends. Like most children on the wired-up side of the digital divide, they go about their day with devices in their pockets or rucksacks, those machines equipped with various child-management and tracking systems.

The parental surveillance and the freedoms the child enjoys are intimately bound together in a quid pro quo relationship. The child's weekly screen time may be charted and is perhaps inversely correlated with the amount of pocket money: the less screen time, the more cash. When the child doesn't immediately answer the par-ent's phone call or text message, the parent pings the phone using the device tracker. When the location where the child's device icon hovers isn't the same place as the child said they'd be, questions and consequences follow. If a friend cracks an off-colour sexual joke or sends a rude picture on a messaging app, the knock on the bedroom door or the arrival of the text from mum or dad comes about a minute later. *Hey. Can we talk?*

One day the child realises that there is at least one form of tracking the parents are doing that the child hadn't known about – a camera in the child's bedroom, part of the household's video-security system. The child tells their parents it feels 'creepy', that they don't like it. *Tough beans*, they're told. *You're ten. We need to make sure you're safe when we're at work.*

The experiences this child is having will feed the framework of how they understand and view themselves, other people and the world. Particular incidents or patterns of occurrences that happen in our childhoods make strong emotional, cognitive and even phys-ical impressions on us that can last a lifetime. When I ask my adult psychotherapy clients about the rigid beliefs and assumptions that cause them the most trouble today, they might not always recall their earliest origins, but they invariably link them to things that happened in their school-age years.

These bundles of thoughts, feelings, images and sensations are the 'early maladaptive schemas' (EMSs) referred to in the last chapter, which are forged in infancy and toddlerhood and embroidered on at this stage of development.[19] If you think of yourself as lazy, hardworking, bold, fearful, autonomous or clingy today, you can probably identify the building blocks that went into that self-concept from your memories of the school years and can probably also recall things your parents did or said during that time that helped shape how you still see yourself now.

EMSs are complex beasts, and most of us have more than one, so psychotherapists and psychologists sometimes use measures to identify a client's strongest. One questionnaire focuses on the EMSs themselves,[20] and another uncovers how the behaviour of the client's parents might have contributed to their formation.[21] *My parent worried excessively that I would get hurt,* says one item, asking the person to rate the statement on a scale of 'completely untrue' to 'describes my parent perfectly'. *My parent overprotected me. My parent made me feel I couldn't rely on my decisions or judgement. My parent did too many things for me instead of letting me do things on my own. My parent controlled my life so that I had little freedom or choice.*

Looking at these statements, imagine how strongly the Gen Alpha child described above might endorse them, and how the parents' well-intentioned use of technology might have shaped that child's psychological development at precisely the time that independence, performance and approval are such key themes for a kid.

One core need of childhood, for example, is being able to express oneself without fearing negative reactions. Adults have their own fears and issues, and sometimes struggle with being open and non-judgemental about their child's communications both online and off. At this age, when self-confidence is lower and parental approval is still so central, the gaze of parents can have a significant *chilling effect* – a phrase that usually refers to freedom of speech and assembly being deterred by political or social authorities, but which applies here too. The child may be afraid of exposure; if their parent finds something they don't like on a smartphone or tablet, maybe their parents will see them as a bad kid. They may have desires that feel unacceptable, or they're ashamed of something about their body, or they're filled with awkwardness, convinced that they're stupid or unlovable. Normal childhood and tween territory, perhaps, but problematic when hardened into a schema that lasts into adulthood.

The hyper-connectedness of today's school-age children means that their parents can conduct a daily, comprehensive review of both their child's external and internal worlds, the latter inferred through online conversations, websites visited and search histories. Authoritarian parents, who focus on maintaining control and establishing their parental authority by whatever means necessary, are perhaps the likeliest to react negatively if they find things they don't like. Their kids may face an especially difficult dilemma: do they freely express themselves and talk with their peers online, risking shame, exposure and perhaps punishment, or do they suppress communication and exploration in a quest for parental approval and acceptance?

The assumptions a child makes about the intentions that underlie a parent's actions can be key to their experience being positive or negative. From a child's vantage point, oversight can be absorbed as a message that they're not capable, not trustworthy, not granted freedom because they can't handle or don't deserve it. As a result, they may feel insecure about their ability to look after themselves without a lot of input from other people or without safety nets. My original intention in GPS-tracking my child was keeping her safe from harm, but over time the scope of my monitoring expanded. Suspecting the consumption of unauthorised junk food, I would call her if I noticed her icon hovering at McDonald's, embarrassing her in front of her friends. If I spotted her travelling west instead of east on the train, I'd message her immediately, depriving her of the chance to course correct on her own.

I want you to trust me, she complained when she got home. *I do trust you*, I replied, but that rang true to neither of us. *I don't trust other people*, I then said, telling her scary stories that probably strongly influenced her beliefs about her vulnerability in the world.

When a parent surveils a child, the last thing they usually intend to do is convey to that child that they don't trust or believe in their competence. On the contrary, they may feel – as I did – that I was *increasing* the scope for my daughter to develop independence, in a safer way. In practice, that might have been somewhat true, but other things were going on as well. The minute she deviated from the plan, went a different way, or turned up at an unexpected location on the tracking app, I was on the case – not judging, I thought, just querying and checking. In the process, I undermined her meeting of another core emotional need, spontaneity and play, including the kind that's safely naughty, like making a detour to McDonald's with your new friends from school.

Some children like and welcome parental surveillance, feeling more secure knowing mum or dad is watching and knowing where they are. But other Gen Alpha kids are pushing back. Seeing their parents' concerns and surveillance activities as overweening and limiting, they might respond with a reactive surfeit of confidence – *I don't need anyone's help, I can get around just fine* – and not exercise enough healthy caution. They might sneak around, rebel, constantly find ways to evade parental surveillance. As childhood gives way to adulthood, both a tendency to throw all caution to the wind and a vulnerability to crippling insecurity can cause problems for newly fledged adults. I wonder if the scaremongering tales I once told my daughter about the world's dangers transferred my own fears into her, implanted the idea that danger is always lurking around the corner, online and off. In myriad ways, technology is both the fodder for this anxiety and a primary pillar of a doomed strategy for assuaging it.

Of course, parenting also comes with insecurity and anxiety as standard features, which is the very reason apps such as Life360 are so popular, why most mums and dads monitor their children using technology – again, 15 per cent 'almost constantly'.[22] Before child-monitoring tech became so advanced, parents had to cope with their kids being out there in the world, independent, apart – there was nothing you could do, and you got used to it. The anxiety extinguished, or at least diminished.

Now, we stoke the flames of insecurity. Surveillance technology, despite what's advertised, doesn't take the fear away – it does the opposite. It reinforces it, keeps it alive. You worry, you consult the app, you see the icon or the text, you feel momentarily better, and the app-consulting behaviour is reinforced – we can become problematically dependent on the checking for a sense or illusion of safety. But we need to do it in today's world, we tell ourselves. We impress on our children, again and again, that there are lots of dangers, that we love them and that we are using this technology only to keep them safe. This is what they learn. No wonder the whole of society now limps under the weight of its constant vigilance to potential harm.

Children of school age gravitate away from parents towards peers partly because child-to-child interactions are so refreshingly non-hierarchical, more equal, sometimes less intense. For a child who's spreading their wings, this time spent in more egalitarian relationships is essential. However much their parents love them, children see themselves as recipients of adult actions and – where things are going wrong – the victims of them.

Before the advent of technological monitoring done at a remove, caring about and paying attention to your kids' activities was associated with greater self-esteem in your child. The extent to which parental monitoring is psychologically healthy for kids depends on whether it's accompanied by true emotional support; whether there are open, honest interchanges between parent and child; and, importantly, on the child's balance of internal vs external control. An internal locus of control means more core emotional needs are being met: they feel competent, have a sense of autonomy and freedom, and are confident in themselves and in their self-control. An external locus of control, on the other hand, leaves the child feeling helpless, dependent and insecure, like a straw in the wind. The adults are always taking matters out of the kid's hands, issuing directives, sweeping in to rescue or reprimand. Parents are in the driver's seat. They're holding the phone.

There have always been protective parents, strict parents, nosy parents, but they never had this much firepower, this much capability. And when you have the power in your hands, it's so very, very hard not to use it.

When a global pandemic descended, children's dependence on technology deepened. Sociologist Michael Wesch was among the first to use the term 'context collapse', referring to multiple audiences flattening into a single place.[23] During lockdowns, millions of students and families experienced just such a context collapse: pupils were forced, often using remote learning,[24] to be simultaneously present at home and school.[25] Walking into post-lockdown classrooms, students were greeted with a far more sophisticated and embedded technological infrastructure. School had changed forever.[26]

The monitoring generally starts as soon as students enter the grounds: pupils in British schools are surveilled as much as prison inmates or airport travellers.[27] Arguing legal basis based on safety and security for pupils, staff and premises, any primary or secondary school in the UK may equip their buildings and grounds with cameras; monitor students' school devices and internet usage with tracking software; follow students' movements using chips in school IDs; and even install biometric scanners at entry and exit points or in the canteen to access free school meals.[28] In a 2022 report on biometrics in schools, the UK Biometrics and Surveillance Camera Commissioner, Fraser Sampson, argued that these technologies were being implemented rapidly and without due care and consideration.[29] But once students are settled into their seats and ready to learn, the datapoints come from EdTech.

The Western media have reported with cynicism and alarm on the sophisticated, AI-powered EdTech kit currently being used in schools across China, courtesy of billions of dollars of government investment. On the headbands that students dutifully don at the start of the day, brainwave measurements taken at ten-minute intervals generate a red light that means the child is concentrating, or a white light that means focus is drifting – although perhaps the pupil is just fidgeting, or the headband isn't tight enough. Surveillance cameras and cute robots scan the classroom, analysing children's health and counting how often they yawn. The data for every child in the class are beamed to the teacher's computer for real-time monitoring; to a chat group where parents can compare their own child's performance against their peers and respond accordingly; and to the government for purported research and development.[30]

China's EdTech has provoked significant concern and commentary among proponents of freedom and democracy worldwide, particularly because of how these children's data might feed into China's developing 'social credit' system. In a centralised social credit situation, the state monitors diverse aspects of individuals' online and offline behaviours, allocating perks for model citizens and removing privileges from those judged not to meet strict moral and social standards.[31] 'The West should not copy any aspect of social credit,' says Samantha Hoffman, an expert on the Chinese Communist Party's approach to state security.[32] 'There is nothing any liberal democratic society should even think about copying in the social credit system.'[33] Indeed, the concept of socially scoring or ranking anyone, much less young people, might make you shudder. But if you're a parent or educator, are you familiar with the popular educational technology tools Google Classroom and ClassDojo?

Statistics indicate you probably are. An astounding 95 per cent of US schools use ClassDojo from kindergarten through to eighth grade, the company claims, and the UK is among 180 other countries that deploy it.[34] A 2022 report from the Digital Futures Commission/5Rights Foundation reported that the platform is widely used in the UK as a space for students to upload work, a place where parents can see their own and other children's 'Dojo Points'.[35] These are awarded for behaviours judged by the classroom teacher to be good, including helping others, staying on task, completing homework, participating in teamwork and demonstrating persistence; they're subtracted for failing to submit homework, speaking out of turn, being unprepared, bullying and disrespect.

I'd never heard of ClassDojo until my seven-year-old asked if she could download it onto the family iPad. She wanted to monitor the behaviour of her cohort of Beanie Boos, small stuffed toys with big glittery eyes. On further questioning, she revealed that her points went up and down on a screen projected on the wall for most of the school day, and she fancied turning the tables when she got home. The school had requested our permission for field trips, religious and sexual education, and use of our child's photograph in promotional material, but they never mentioned using EdTech to track classroom behaviour. If they had, I'm not sure what we could have said about it. Among the issues with EdTech identified by the Digital Futures Commission were significant problems with consent. Power imbalances and social pressures make it hard for an individual family or child to refuse, and giving permission doesn't mean much anyway when you've got no real idea what you're consenting to.

Not fully grasping the potential consequences of EdTech used at their child's school is completely understandable; it's probable that the school doesn't really get them or can't fully control them either. One problem is that children can easily be led into less safe digital waters: both Google Classroom and ClassDojo both serve as jumping-off points to outside apps and platforms with different levels of data security. Another is what EdTech apps can technically get away with through opaque and confusing privacy policies, which may still allow platforms to collect and sell unknown quantities and types of data about young learners, drawing on extraordinarily rich stores of information about children.[36] Yet another is the hacking danger: in 2018 the Federal Bureau of Investigation (FBI) issued a warning about EdTech, highlighting the sensitivity of data commonly stored, including personally identifiable information, biometrics, academics, behaviour, health, web browsing, geolocation and classroom activities. The FBI reported that cyber criminals were targeting schools and using data to shame, threaten and bully children; after one large EdTech company suffered a breach, student data went on sale on the Dark Web.[37] In early 2023, an organised hacking group stole highly confidential pupil data from 14 schools across the UK.[38]

Some platforms have made improvements – Google Classroom passed its last privacy check-up by the Common Sense Privacy Program in late 2022 – but ClassDojo failed, and new apps keep proliferating.[39] The Data Future Commission report concluded that managing how children's data are processed and used is a 'near-impossible burden on any school, parent, caregiver or child'.[40] So, should we just shrug our shoulders at a situation that's seemingly spun so far out of our control?

Maybe not completely – parents, teachers, schools and regulators still retain some influence. After the early-pandemic, necessity-driven rush to adopt and implement all this EdTech, maybe it's time to re-evaluate what's truly needed, to sort the wheat from the chaff, and to consider seriously the damaging psychological effects of social scoring in the classroom for kids trying to find their feet in the world.

In Western classrooms, data on pupils aren't (yet) being collected from brainwave-measuring headbands and face-scanning robots; instead, what goes into the system are inputs from individuals. ClassDojo has a variety of functions, but its original and enduring application is subjective, teacher-generated perceptions and judgements about how students are acting: focused or fidgety, angelic or impish. 'These human judgements, which may or may not be fair or biased, are manually entered as facts into the app or website as behavioural (and arguably biometric) data,' write the authors of the Digital Futures Commission. They propose that this is no better than the practices of authoritarian governments, and point out that the European Economic and Social Committee (EESC), in a draft

European Union legislation on AI, has proposed an all-out ban on social scoring of this kind. So, why are schools doing it?

To understand more, I call on Laura,* head teacher at a London primary school.[41] We sit in her kitchen, and our respective children – all of whom experienced ClassDojo at their own primaries – hang out in another room. Laura doesn't use the app at her school, but says that if she'd wanted to, she might not even have thought to consult the school governors or request the permission of parents. She's a switched-on senior educator of long experience, but our conversation is the first time she's ever considered the wider privacy implications of the ClassDojo platform described above.

Laura says she understands why teachers are drawn to apps like this, wondering if they particularly appeal to less experienced or more overwhelmed educators, concerned about how it might reflect on them if their class is underperforming. Plus, the cartoon-monster avatars and gamified aspects of the app can be hard to resist, especially when they keep many children motivated, particularly those who are more competitive and achievement-driven anyway. 'What really sold the programme to me was the effect it had on students,' reported one educator in the *Guardian*.[42] In a blog post about 'How ClassDojo Helped [Her] Become Teacher of the Year', North Carolina primary school teacher Jachie Pohl described how the app 'we all know and love' facilitated quick communication with parents.[43] 'Easily stay connected with your child's classrooms on ClassDojo!' says a template the platform provides as an info sheet for parents. 'You can see all of your child's feedback from teachers, hear important announcements and updates, and see photos and videos from class!'[44] Pohl also appreciated how the app helped

---

* Laura is not her real name.

her reinforce the 'moral focus virtues' at her school, which include Wisdom, Gratitude and Self-Control.

So, if it's so helpful for teachers and they love it, potential privacy issues aside, is there an issue or immediate negative impact? Isn't ClassDojo just an improvement on the old-style posterboard 'traffic light' behaviour charts children have had in classrooms for years?

Laura explained that neither method is ideal: in terms of both motivation for learning and healthy identity-building, ClassDojo is simply a glossier, techier version of behaviour-management methods that had been falling out of favour because of their recognised impact on children's self-esteem, particularly those kids who were already operating with challenges. Whether it's a paper traffic-light poster or a fancy app, when children's names or photos publicly plunge from green into amber or red, or their monster constantly languishes at the bottom of the ClassDojo league table, it can evoke embarrassment and shame. Some children might be able to take the hit, using it as an incentive to change, but others might convert their feelings of inferiority into angry, disengaged or outright rebellious behaviour, creating a confidence-destroying downward cycle. 'It's teaching a child that their way of behaving isn't acceptable to society,' Laura says, 'and that *they* aren't acceptable.'

The direct effects of social-scoring points systems in the classroom are that they reward students' conformity with rigid standards of behaviour. That may result in an apparent win – a quieter, more biddable classroom – but it doesn't necessarily teach pupils to navigate their lives confidently and competently outside and beyond school. When teachers focus on punishing and rewarding specific actions and outcomes, Laura says, they're not encouraging a growth mindset, communicating to kids that it's okay to make mistakes, struggle with difficult things and improve with time.

Behavioural management systems like this have also been criticised as ableist, disadvantaging children whose fidgeting stems from attention deficit disorders, whose social difficulties are related to autistic spectrum issues, or whose apparent naughtiness is a communication of distress. This acknowledgement brought Laura and me onto the topic of automated communication with parents via EdTech. Although this ClassDojo feature doesn't have to be enabled, it can be set up for parents to receive push notifications about their child's good and bad behaviours or academic achievements. Mum or Dad might get a midday boost that Annie has once again received a point for her writing, or be plunged into despair at the news that, for the tenth time this week, Calum has hit his table partner in the head.

When notifications pop up on our phones, it's a nudge to *do* something; we are conditioned to react and respond. Action might seem even more urgently called for when the ping is something to do with your child. But what's the parent to do? What's the action or reaction meant to be? Perhaps a teacher hopes that parental intervention or discipline at home will reduce the time or effort needed to address a problem within the time- or resource-poor school day, but constantly notified parents could become a victim of a type of 'collateral intrusion' – their privacy, freedom and focus might be affected by monitoring of which they themselves are not the target. 'The parent of a neurodivergent child, for example, is probably having all sorts of challenges,' says Laura. 'If every day you're receiving nine pings from school, here's an issue, here's an issue, and you're receiving them in live time, at work trying to do your job, how are you supposed to function?' And, on the children's side, says one commentator, don't kids 'deserve privacy, personal space, and a learning environment where their every transgression is not reported back to their guardians'?[45]

Laura also worries about another dimension of transgression reporting. When one of her pupils gets into trouble, Laura often wonders what will happen when that child's carers hear about it. Her primary school is in a socioeconomically deprived, high-crime area, where many students live in difficult circumstances. Familiar as she is with her school community, when Laura reports a child's infraction to a parent in person or on the phone, she knows there's a possibility that the punishment the child receives at home could result in a safeguarding issue. That knowledge requires careful handling of the situation, and Laura's not about to hand that responsibility over to an app. The Digital Futures Commission report underscores this problem as well: where children are affected by living in homes where they already experience violence, it says, constant behavioural reports can make their lives worse.

Finally, Laura mentions the unfortunate ongoing realities of teacher bias, which persist even after education and continuing professional development that's designed to raise awareness and reduce the impact of that bias. Julie Garlen, an Associate Professor of Childhood and Youth Studies, has argued that ClassDojo's avatars can impede recognition of implicit bias by 'eras[ing] individual differences, including race, religion, physical abilities or gender identities',[46] but that's not what Laura's most concerned about. She worries more about the individual teacher's explicit and implicit biases, prejudices and assumptions influencing their awarding and deduction of points, transforming inevitably subjective observations of pupil behaviour into a permanent educational record that might look to others like fact.

I express amazement that problematic levels of teacher bias could be common even in a multicultural city like London, so we call the three kids into the room. First, we ask them about ClassDojo, and one of the children – a confident type who's just emerging from the Industry vs Inferiority stage with a strong sense of competence – can't see that ClassDojo is anything other than a neutral, transparent meritocracy. 'A person does something good, they get a point,' she shrugs. 'They do something bad; it's taken away. It's not a big deal. It's not that embarrassing.'

But then Laura asks a different question: 'Who's better behaved in class – boys or girls?'

Girls, they agree.

'And who is most likely to get in trouble?' Laura asks.

'In my class,' another girl pipes up matter-of-factly, 'it's always the Black boys who get into trouble. Yeah. Always. No one else gets in trouble except them.'

'Why do you think that is?' Laura asks.

'They don't do anything,' the girl continues, shrugging. 'They just get blamed for it. I talk a lot in class, and I get a warning, not a detention. They talk once and get a 40-minute detention, and they go back to the same lesson the next day, and the teacher expects that from them again.'

They dash out of the room to do whatever they'd been doing when we interrupted them.

'From the mouths of babes,' Laura says, when they're out of earshot, and indeed, Black Caribbean children, particularly boys, are far more likely to be excluded from school than White British children.[47] 'This bias is everywhere in education,' Laura sighs. 'We're only beginning to scratch the surface of it.' She doesn't have behaviour-management apps in her school because she doesn't want teacher bias transmogrifying into datasets that then carry an illusion of truth or accuracy, especially when that data might have long-term impact. Sometimes that means she doesn't have the data that inspectors or governors sometimes ask for ready to hand, and she's okay with that.

But hers is just one school, and conscientious objectors like Laura seem to be in an increasing minority, even as the Digital Futures Commission's impressive panel of experts bangs the warning drum. Their report paints a picture of a future in which universities, employers and even insurers can instantly access young adults' 'full life and development' files; can review their academic history, behavioural data, attention span, failures and strengths from primary and secondary school; and can make decisions based purely on that information, without the person in question having any recourse.[48]

In other words, there's every possibility that today's EdTech, including its social scoring, will feed tomorrow's algorithms of oppression.[49]

'What's the great pay-off that makes that acceptable?' asks Laura. 'You don't need it. Good education . . . is based on the good-old-fashioned solid relationship between the teacher and the pupils in their class. They don't have to love you, but they have to respect you, and that respect comes from feeling safe. Safety doesn't come from *I can add a point or take it away at my will.*' Confidence and courage come from a kid feeling, she explains, as though they're still worthwhile even though they're different or they goof up, and that tomorrow is another day.

'Without that fresh start at the school gates every morning,' she says, 'a kid spirals.'

Experiences in the classroom can either encourage a student's confidence or cement their belief that they're inferior, and the consequences – including spiralling – can extend well beyond the school gates. Professor Pete Fussey, an expert in surveillance and the new field of algorithmic justice, believes we should be concerned about any situation in which children – among the most scrutinised group of people in society – experience yet more intrusions on their privacy, whether through educational surveillance or facial-recognition technology in public spaces. Although we'd like to trust the government and police deploying this surveillance to be fair and impartial, he says, we can't expect perfect self-awareness and behaviour from the humans that make up these institutions, any more than we can expect it from teachers.

'Both trust and surveillance are unevenly distributed in society,' Pete explains.[50] 'It's going to be the social groups on the margins of society that experience more intense and more punitive surveillance than others.' And certain social groups are clearly on the margins. For example, the statistics for juvenile stop-and-searches in the UK mirror the trends for school exclusion: while Black children make up only 4 per cent of the population of ten- to seventeen-year-olds, they account for 18 per cent of stop-and-searches, 15 per cent of arrests and 29 per cent of the youth-custody population. White children reoffend at a rate of 35.3 per cent, but for Black children the rate is 42.4 per cent.[51]

On the afternoon that Pete and I speak, he told me he'd spent that very morning on a police ethics panel, discussing the force's data analytics capability. Among the topics of conversation that day had been intelligence dossiers on children as young as nine. The reason for the authorities holding such dossiers, Pete says, is the presumption that these kids will continue to commit crimes in the future. Once the data from these dossiers have been enfolded into algorithms that have the illusion of clean mechanical objectivity, he says, we lose sight of the prejudiced and flawed information that trained them and inputted the information. 'There are certain forms of knowledge that we give more credibility to than others,' Pete says. 'If something is technical or technological, it has a kind of authority that's not necessarily warranted.'

The data points that go into making predictions about a child's future, when they're stripped of their context and filtered through a series of all-too-human biases, can become a self-fulfilling prophecy. The same kids who are most closely watched in school, from whom no good is expected, are the kids and young adults selectively stopped and searched,[52] or the ones who are arrested because facial-recognition technology (FRT), using algorithms that have been developed using white faces, is comparatively inaccurate when distinguishing individuals of colour.[53]

And when mistakes are made, justice systems can find them hard to admit. In 2020, Robert Julian-Borchak Williams was arrested in front of his wife and two young daughters for allegedly shoplifting watches. FRT had supposedly identified him from a surveillance video, but he hadn't been inside the store for six years. The detectives questioning him quickly realised that he was not the man in the photograph. 'I guess the computer got it wrong,' admitted a detective. Still, Mr Williams was not immediately allowed to go. He was released on bond and later required to appear in court.

An entirely innocent man, perceived as guilty only because of a flawed FRT match, this African American man still felt ashamed. 'It's not something I'm proud of,' he told the *New York Times*. 'It's humiliating.'[54] Somehow, an unfair and false accusation had changed how he felt as a person. His children were watching closely, trying to interpret what had happened and why, and perhaps forming long-lasting beliefs and assumptions about how people who look like them can expect to fare in society. After his arrest, his daughter took to playing cops and robbers, accusing her father of stealing things and locking him up in the front room.[55]

Biased humans may direct their attention unequally, but there's as much of a problem when the technology is pointed apparently dispassionately at *everyone* – like CCTV cameras at a public event, or cameras lining a main road. FRT, developing faster than regulation can rise to stem the tide, is something that young people mistrust – a point my nine-year-old might have been driving home when she smushed that overripe banana into the watching eye of the video doorbell.

Pete Fussey explains that even when surveillance is done remotely or discreetly, it communicates implicit messages about power, citizenship, trust and freedom. How you decode and interpret those messages depends on who you are – where you're 'socially located', he says, a phrase that refers to a combination of your gender, race, social class, age, ability, religion, sexual orientation and geographic location.[56]

'I'm a white middle-class academic in my mid-forties,' he says. 'Being surveilled doesn't really matter to me. But if you're the pupil in the referral unit who's always getting shit from authority, then surveillance communicates something different.' I remember the white middle-class pre-teen shrugging, saying it didn't matter if her behaviour were tracked or she were docked points in class; the privilege of her social location means that she trusts everything's probably going to work out okay in the end. A different kind of child, suffering from insecurity and inferiority and struggling with their life and identity in a society that feels harsh, might experience monitoring differently.

But even the most joyful moments of young lives now fall within the widening circle of childhood surveillance. In 2020, *The Washington Post* reported that facial-recognition services were being rolled out in summer camps around the United States, enabling doting parents to monitor their offsprings' fun levels from a distance.[57] To date, I've had no idea if my child was having a good time at camp – if she were homesick, or having trouble making friends, or feeling unwell. If she were having a joyous time at camp, I couldn't share in that experience either.

I imagine the two rows of pine trees lining the drive to her camp, an iconic sight for all who love it there, kids who cannot contain their excitement at being on the cusp of total freedom from the everyday, from parental oversight and the demands of their usual lives. I picture a watching eye on every tree, nailed to the rugged fragrant bark, peeking from between the long green needles. If that happens, once that happens, once we become accustomed to it – how long will it take us to remember, if ever we knew, what problem we were trying to solve?

No one escapes the school-age years without picking up a few lasting maladaptive cognitive frameworks. My clinic is full of people suffering from perfectionism, unhelpfully rigid beliefs about success and failure, and the tendency to conform to what they 'should' do rather than what makes them happy. So many of them grapple with an 'unrelenting standards' schema; they're their own harshest critics, although they also worry endlessly about how others will judge them. Exploring these problems together, we'll often find that they're a bit developmentally stuck: they never emerged with competence and confidence from the Industry vs Inferiority stage. Part of them is still back in their schooldays, still uncertain whether they can make it on their own.

I don't know whether I'll continue my clinical career long enough to work with adult Gen Alphas, but their childhoods have been one big social experiment, and the results aren't in yet. I can only imagine that spending one's formative years as a ward among wardens, surrounded by tech that allows parents and their proxies constantly to keep tabs on one's movements, communications and explorations, has got to have an impact. Worst case, I imagine a future cohort of young adults who don't trust themselves or the world, whose young identities were forged in fear, doubt and insecurities about their safety, their privacy and the judgement of others. A generation of well-meaning and digitally hypervigilant parents might be producing a glut of chronically anxious, stressed, cognitively drained Alphas.

And then there's the school environment, a major player in the normalisation of surveillance for today's youngest generation. The people I see in clinic are already plagued with competitiveness, occupational stress and the scourges of social comparison and imposter syndrome, and they grew up in relatively analogue classroom environments. I shudder to think how an educational system steeped in social-scoring behavioural-management platforms might deepen these problems or make them more endemic.

I hope I'm wrong about Gen Alpha, and, in any case, I hope we'll have figured things out by the time Gen Beta starts arriving in 2025. I wish I had quick and easy answers for parents, but I know from personal experience how hard it is to balance a child's developmental need for experimentation and autonomy with parents' concerns about safety. If you're a parent, for every electronic-surveillance decision you make about your kid, ask yourself some searching questions. What's convinced you that this technological solution is necessary? What clever marketing, groupthink, scaremongering headline, or anxiety might have compelled you? Assess and perhaps research the whole picture: do you reckon this app or gadget is *really* going to decrease your worry? If you don't choose to go ahead, what would the alternatives be? If you do proceed, what explicit or implicit messages might your kid be receiving from you, and how might you balance out some of the more negative ones?

I chose Confidence vs Insecurity as the new developmental challenge for this age because Erikson's Industry vs Inferiority felt too confined to the educational setting, and tech's ubiquity means children are *constantly* in danger of experiencing the confidence-shaking, insecurity-stoking psychological effects of blanket surveillance wherever they are: school, home or out and about. Thanks to EdTech, school's a major issue, but the research, reviews and expert reports emerging now may be causing more schools to question the uncritical implementation of social-scoring platforms like ClassDojo, and indeed any tech that could compromise a child's future rights and prospects. If education regulators and school boards start reflecting more deeply, that will be a good thing.

Parents have a role to play as well. They can make their own lives more manageable and foster an arguably healthier degree of separation between their child's home and school lives by opting out of automatic notifications. At home, they can encourage kids to accept themselves as they are, celebrate their differences, understand their unique gifts and learning styles. They can ask teachers, schools and governing boards tough questions about any behaviour-tracking technology used in the classroom, raise the problem of bias and point out the potential pitfalls of the tech.

We live in an age where technological solutions to all sorts of problems are championed and reified, and many people are blinded by shiny and amazing innovations. When you're querying or protesting something as innocuous-sounding as a classroom resource that manages behaviour and facilitates parent–teacher communication, it can be hard to go it alone, but there's strength in numbers. Has the material in this chapter made an impression on you? Well, it just might make someone else sit up and take notice too.

5

Adolescence

Jessie's adjustment to secondary school was hard, sometimes heartbreakingly so.* Her classmates were performing dances she didn't know, giggling over memes she hadn't seen.

Jessie's mother was torn as she watched from afar. Would she have to choose between Jessie being either safe or happy? With her heart in her mouth, she finally said yes. Just like that, her barely adolescent Generation Alpha child stepped into an alternate universe that her parents could track but not fully understand. Jessie had joined TikTok.

Jessie's initial 'For You' landing page was the app's best first guess, composed of what was trending with TikTok's 1 billion monthly active users.[1] Supplied with only this sparsest of demographic information – Jessie's location, gender, age – the algorithm sat and waited to be fed.

In the first nine months of 2020, more than 100 million households globally streamed at least one episode or film of Japanese animation on Netflix, a 50 per cent increase on the previous year.[2] In 2021, the pandemic-era release of *Demon Slayer*, an anime film, netted £68 million outside Japan.[3] Since then, at least for a period, walking down the street without spotting at least one teenager in an anime T-shirt became statistically improbable.

---

* Jessie is a pseudonym.

Jessie was an emergent member of this tribe. The first T-shirt she bought with her own money depicted *Demon Slayer* heroine Nezuko Kamado, a young girl-demon possessed by feral frenzies so dangerous she wears a bamboo muzzle to protect the populace. Perhaps, mused Jessie's mother, Nezuko was a metaphor for the storm of female puberty.

Anime's burgeoning popularity in 2021 ensured that someone dressed as Nezuko was nestled among that initial collage of TikTok videos. In a flush of recognition and appreciation, Jessie clicked the heart icon beneath. Like a snowball rolling downhill and contemplating a career move to avalanche, she was rapidly and willingly pulled in and along by what felt like a force of nature.

'I was surprised how liking one video could recommend me so many more,' Jessie said. 'I knew they would take what you like and maybe give you a few videos, but not *that* many.'[4]

So, so many. Jessie rapidly formed a mental map of all the subcultures within the anime community, following trajectories that took her both further into and right out of TikTok. She counts off the various branches of the anime-fan family tree, on fingers capped by nails at least as long and tapered as the Japanese girl-demon's: the editors, the cosplayers, the artists, the skit-makers, the fan-fiction authors.

Jessie aspired to cosplay – dressing up like fictional characters and often sharing the results on social media – but was intimidated by other people's skill. Still, she followed the cosplayers avidly, including one who achieved eerily accurate effects with his art. She described him as a nice person who cares about how his viewers are doing.

'Going on his TikTok stream was the highlight of my evening,' she said. 'It made me feel better. I felt like I couldn't explain that to my parents. I didn't think they'd understand. *You don't know this person. Why do you talk to them? That's not a thing you should be doing.'* Jessie was worried her parents would take it away.

One evening her mother confronted her with a screenshot of the amount of time Jessie had spent on TikTok. The new limits were effective immediately, and Jessie panicked. Who could she talk to now? Her parents thought offline friendships were safer than on, but for the tween it felt like the other way around. Unlike the mean girls at school, if the cosplayer hadn't been kind, she could simply have unfollowed and blocked him. With a 15-minute daily limit, Jessie couldn't attend a TikTok two-hour livestream. But on her phone, she still had a photo of the cosplayer who supported her on a hard day.

The wig is grey and shaggy. Reddish-brown abrasions or lacerations cover the face and neck. Behind the fringe, the eyes look black and blue, but crinkled in a friendly way, and the character's smile is broad and happy.

This excoriated character is Tomura Shigaraki, Jessie said, from *My Hero Academia*.[5] She didn't care for the film but liked the image. PTSD makes Tomura scratch himself, she said; when he first gained the ability to explode people into dust at the tap of a finger, he annihilated his family.

'He didn't realise what he was doing,' Jessie explained. 'He accidentally touched his mother. She was reaching out to him as she was crumbling. I thought that was kind of sad.'

I'm struggling to keep my face composed, forcing my body into the language of dispassionate interest. But my throat contracts as she describes how Tomura hurt his mother without meaning to, because Jessie – not her real name – is *my* daughter.

I reach for my beloved digital-age child, whose struggles I cannot always know, often cannot fix, and sometimes have unknowingly caused. She is trying so hard to find her way. The love I have for her has given her the power to destroy me, and I can feel myself crumbling.

Ah, teenagers. I would have felt remiss if I hadn't revealed my bias when it comes to this chapter, so I hope having disclosed this increases rather than decreases your trust in me as an author. To employ an overused term, I'm finding it *triggering* to be writing about modern adolescents while simultaneously navigating the teen angst and technology overload roiling my own household. I'm not alone, I'm sure.

Every generation of adults moans that 'kids these days' are nothing like they were in their own youth – and with the ascent of Generation Alpha teens, born from 2010 onwards, befuddled older folks are probably more accurate in their perception than complaining parents have ever been. Today's teens *are* different, thanks to living through the biggest force to upend society since the invention of the printing press. Still, while the context might be different, the biological and hormonal realities of adolescence – and many of the psychological and emotional ones too – remain constant.

Sturm und Drang was an eighteenth-century literary and artistic movement in Germany, a backlash against the boring, strait-laced rationalism of the Enlightenment.[6] Forget reason, forget logic! A sensual, impetuous, intense life was where it was at.

German polymath Johann Wolfgang von Goethe helped launch this emotionally incontinent movement with the first proper *Sturm und Drang* novel, *The Sorrows of Young Werther*.[7] The word 'emo' might have entered the lexicon only around 1985, but Goethe's youthful protagonist was angsty and overwrought two centuries before that.[8] Werther's passions were all over the place, and some pundits have declared he must have had bipolar disorder, but perhaps he was just a teen.[9] Werther's doomed love for the 1700s' equivalent of his high-school girlfriend, Charlotte, ends in his death.

'Werther fever' went pre-digitally viral. Youth across Europe donned custard-yellow trousers and electric-blue jackets,[10] but the more troubling trend was reported copycat suicides amongst moody young aristocrats.[11] Teenagers getting drawn into self-harm by social contagion, a phenomenon whose incidence is much feared and possibly overestimated in the social media age, is still called 'the Werther effect'.[12]

When an American psychologist and developmental theorist started writing a book about teenagers, the name of that old German cultural movement came in handy.[13] The founder of the American Psychological Association, G. Stanley Hall, often translated German-language concepts into an American context: in 1909, it was he who invited Sigmund Freud and Carl Jung to the US, amplifying the ideas and influence of psychoanalysis, and changing the course of our collective social and psychological history in the process.[14] Casting about for a term that captured the perilous transition between childhood and adulthood, Hall hit on the English translation of *Sturm und Drang*: 'storm and stress'. The storm of decreased cognitive control, combined with the stress of a teen's increased social and emotional sensitivity, results in fights with parents, moodiness and the kind of crazy, risky behaviours that come from craving stimulation but not having the emotional maturity to obtain it sensibly.

A century after Hall, technologies including functional MRI scans have given us greater insight into the neurological and biological drivers of teen angst, reactivity and risk-taking. Around puberty, the biological socio-emotional network becomes highly active and reward-seeking, but the cognitive control network – responsible for reasoning, planning and impulse control – matures later.[15] Until these systems come into better balance as the adolescent moves towards adulthood, it can be a wild ride.

Hall's ideas about stormy, stressful adolescence hold up today, and while Erikson's life-cycle model is entirely about identity, the adolescent phase is the only one where the word appears: Identity vs Role Confusion.[16] Erikson said that the way teenagers behave is kids experimenting in the way they need to do, *must* do, to form a strong identity that will serve them well in adulthood. The Earth's mature landscapes were formed through violent upheaval – asteroid strikes, exploding volcanos, relentless glaciers – and teenage angst, as challenging as it is to go through, has a similarly constructive purpose.

Identity continues to evolve throughout our lives, but individuals who have navigated the storms of the adolescent years arrive at some more solid-feeling land by the time they reach twenty, or perhaps a bit later, the years between eighteen and twenty-five often being referred to as *emerging* adulthood.[17] Whatever might yet change, a young person with a psychologically healthy identity has a sense of continuity in their self-image, beliefs, values and desires. They don't know everything, but they trust themselves to figure it out. Feeling strong in these ways gives people on the cusp of adulthood the confidence to commit to initial paths in life: careers, interests, other people. When self-knowledge and self-belief sing in tune with behavioural choices, that's the identity holy grail.

Role confusion, on the other hand, is what the teen years tend to feel like before things coalesce. Countless books and movies portray the familiar teenage tropes – the awkward, confused kid doing and saying cringy things. Fortunate adolescents have family, peers and resources to accept and hold them until they get through the worst of it. But unresolved role confusion that continues on into adulthood can result in the individual's experiencing 'commitment issues' in jobs or personal relationships, drifting from one thing to the next, lacking a clear sense of value or purpose.

The modern teen's technological landscape provides both the backdrop and the vehicle for an identity crisis on steroids. Whether it's performing on TikTok, vlogging on YouTube, creating avatars for gaming, or displaying (and consuming) curated, filtered selves on Instagram, teens can experience the online world as a virtual playground of healthy experimentation, or a nightmarish hellscape of social comparison. Developing a healthy social identity aligned with your inner sense of self can be a fraught business if the online environment is constantly sending you conflicting, confusing or insistent messages about who you *should* be, perhaps pushing you into greater fragmentation: shown and hidden, acceptable and unacceptable, 'living your best life' on social media while suffering inside. Today's digital teens don't merely have to navigate Identity vs Role Confusion – they've got to deal with Harmonisation vs Compartmentalisation as well.

'All the world's a stage, and all the men and women merely players,' Shakespeare famously said,[18] and Erving Goffman, a psychologist who wrote about the social self at the same time that Erikson was mapping the lifespan, took that metaphor and ran with it. In his most well-known work, *The Presentation of Self in Everyday Life*, Goffman argued that the Bard was exactly right: we *are* all actors, perpetually performing on the social stage.[19]

Every time you meet other people, imagine the 'On Air' sign illuminating red. From that moment, you're creating an impression, performing for whoever's watching. When the audience changes, your performance alters too, in imperceptible shifts or wholesale transformations. That's not to say you're pretending or putting up a deceitful false front. Stepping onto a particular stage, confronted with the audience of the moment, you act in a way that will serve you well and show you to best advantage.

The show you put on is partly for that audience's benefit, to make things enjoyable and comfortable for them, but ultimately, it's for you. You prepare and present in ways that will spare you embarrassment or shame, bring in the favourable reviews, and secure or maintain love and approval. When your performance is sincere, it might not be strenuous.

Sometimes, though, it's an effort. The role is outside your comfort zone, but it's the one people want you to play, the one they like; after a while, you reach the point where you can play it convincingly. Soon, people respond to you as though you *are* that role. If you deviate from it, they say *this doesn't seem like you*. Like the actor whose fans assume they're exactly like the characters they play, who are stereotyped by the casting directors, you can become trapped.

But you're not alone. In this never-ending piece of immersive theatre, you're playing to your audience and they're playing back to you. You're sizing up your performance compared with theirs and, whether it looks like it or not, they're doing the same. You could try to evaluate yourself on your own merits, but how can you when 'merit' itself is defined by society? How content you are about yourself mostly depends on how you think you stack up against other people. It's all relative.

Social comparison plays a key role in identity development during adolescence and emerging adulthood, when relationships with people outside one's family of origin become more critical.[20] Comparison is ubiquitous and inevitable because it's old stuff, hard wired into you. The nineteenth-century biologist Herbert Spencer didn't call it survival of the fit, but survival of the *fittest*.[21] The compulsion to be constantly assessing how you're faring relative to the crowd is an embodiment of the mechanics of natural selection. Scratch the surface of any human interaction and you'll expose the fundamental preoccupations and brute machinations of your ancient brain: reproduction and survival. The organism that's best adjusted to its environment will have the best chance at both, and you're an organism like any other.

When our environment became digital, performance and comparison became trickier. In Goffman's day, before the internet, it was easier for a teenager to present completely different faces in various places, showing up in one guise at school and another at home, inhabiting one persona at the club on Saturday and another at church with the family on Sunday. On social media, however, those formerly separate contexts are collapsed into one space, partially divided up but with a *lot* of potential for deliberate or actual overlap among various audiences.

How do you navigate *that*? And you're not playing only to the viewers you intend – one's digital presence extends beyond the audience you know, and who know you. The strangers out there in the dark, beyond the light cast by your stage, are invisible. Still, you might find yourself using their imagined reactions to shape your behaviour, especially if you've been heckled by the peanut gallery before.

In a networked public such as the internet, you enter a digital space prepared to meet your friends, but casual passers-by might see, comment on and criticise you.[22] Furthermore, what you said or how you looked on a particular day doesn't disappear into the ether, unless you're talking Snapchat, which is designed to share images that appear briefly and then vanish into permanent inaccessibility. Outside platforms such as this, networked publics tend to hang onto your interactions and proclamations for a long time. Moments that could have been fleeting have a wide reach and a long shelf life.

No wonder young people and emerging adults, the demographic with the highest frequency of tech-mediated interactions, are so concerned about their presentation of self on social media. Forty-three per cent of teenagers report posting only content that makes them look good to others, and 37 per cent said they feel pressure to share things that they predict will get likes and comments.[23]

Experiences online are likely to have long-lasting implications for an adolescent's sense of self. Such an environment can be unforgiving, as the rise of cancel culture has shown us. Comments or behaviours that were understandable in one context become seen as unforgivable when translated into another. A gaffe or mistake can dog your footsteps for months or come back to haunt you when you least expect it. Years later, thanks to information being replicable and searchable, you can find yourself being judged for a sin you committed out of youthful indiscretion or ignorance, or being held accountable for something a troll or frenemy put out there about you, or an impersonating catfish put out there *as* you.

The phrase *you'll never live this down* predated the internet, but the potential for public humiliation has never been a more plausible anxiety for so many.

The capacity for comparison is infinite online. You're no longer just 'keeping up with the Joneses', weighing yourself against physical neighbours and 'in-real-life' friends. You can compare your looks, success, popularity and happiness levels with millions of rivals around the world, utter strangers performing their best lives on Instagram for your delectation and envy. That could be tough on anyone, including the lucky adults whose social identities are relatively well established, and whose self-esteem, self-worth and support structures are more solid. Imagine, though, what it might be like as a teen growing up in a social-comparison extravaganza, and hence it's been hypothesised and assumed far and wide that social networking sites, especially visually orientated ones such as Instagram, are a 'perfect storm for negative social comparison experiences'.[24]

Unsurprisingly, six decades of social-comparison research show that downward comparisons, in which we rate ourselves as superior, are associated with more positive feelings.[25] However, we naturally skew towards upward comparisons, rating others as doing better than us. Sometimes upward comparisons lead to improved self-concept and helpful actions by inspiring and mobilising us towards positive goals and results, known as the 'assimilation effect'. Generally, however, sizing ourselves up against others and finding ourselves wanting is associated with feeling worse, and teens mired in the psychological and physical changes of adolescence might be particularly prone to feeling worse in any case.

Still, the actual impact of scrolling on Instagram, and whether there *is* a predictable, reducible effect, is tough to determine and more variable than we might imagine. Experimental and longitudinal research that looks at causation rather than correlation has demonstrated that happier teens are relatively less affected by what other people are doing and how they're looking, and more able to *benefit* from upward comparisons and the assimilation effect. Depressed adolescents already suffering from low self-esteem tend to compare themselves unfavourably with others and feel worse after doing so, getting trapped in a negative cycle.[26] *What* is being compared also seems to make a difference. While teens who measure their abilities and achievements against other people's on social media feel more distress about their identity, get caught in more negative thinking and have less clear self-direction, those who explore their developing selves through expressing, comparing and arguing *opinions* online have a greater degree of healthy self-reflection and a stronger, more positive sense of identity.[27]

Such complexities challenge our compulsion to find certain answers to our worries about tech, and those anxieties are understandable. After all, whatever a teen is doing online, however positive or negative their current interactions, it's hard to argue against the notion that the social media environment creates a lot of scope for problems: exponentially increasing the targets for comparison; unhelpfully stripping away context that would show the bigger picture of other people's lives; and creating an environment where behaviours are shaped through being constantly watched, scrutinised and judged – or even punished.

In the eighteenth century, an English philosopher proposed a radical innovation for prisons, a structure that would manage prisoners' behaviour not through violence or physical coercion, but by controlling their minds, 'in a quantity hitherto without example'. For his 'Inspection House', Jeremy Bentham proposed a panopticon: a circular building with the convicts' cells arranged around the perimeter.[28] Standing in the middle would be a tall, all-seeing inspection tower where the overseers sat, invisible to the prisoners but able to peer into each cell and communicate with inmates through 'conversation tubes'. Because they would assume the inspector was always watching, Bentham said, the prisoners would be forced to behave themselves to avoid punishment. Few panoptica were built in his lifetime, but the one in which we're currently imprisoned is a more powerful instrument of control and behavioural regulation than any bricks-and-mortar structure he could have envisioned.

We are all the watchers, and all the watched. Again: that's a struggle for adults, but these poor kids! Developmentally speaking, aren't they supposed to be making mistakes without such a high risk of being shamed, or some worse consequence?

'Trace through the [potential] impact of surveillance on a kid,' Professor Pete Fussey says – you'll remember him as the Facial

Recognition Technology (FRT) expert who's also the director of the UK's Centre for Research into Information, Surveillance and Technology (CRISP). He reminds me that social media platforms aren't the only place where modern teens feel forced into curating and censoring themselves.[29] FRT at events or on university campuses could stifle freedom of assembly and expression, discouraging young people from protest and political action, affecting not just identity formation but human rights.

'When you're a teenager, you *should* experiment with your life,' Pete points out. 'You have to push the boundaries because a lot of the laws being imposed on you *should* be challenged. That's a healthy thing, but we're inculcating this expectation and strong message that self-censorship is really important.'

He tells me about a study on the public acceptability of live facial recognition (LFR), a technology where algorithms attempt to detect, in real time and with questionable accuracy, live CCTV footage of individuals rightly or wrongly deemed suspicious, or people wanted by police.[30] Nearly 40 per cent of teens and emerging adults aged sixteen to twenty-four reported that they'd be reluctant to go to an event in any venue where LFR was being used for purposes like that.

The potential effect of constant online and public surveillance on the developmental needs of teenagers is, in multiple senses of the word, rather chilling. Are the kids all right? *Can* they be? For those who dislike and question the situation they're in, resistance seems futile, not an optimal scenario for a young person who is finding their own power, who is developing their own voice. Moreover, for disenfranchised and vulnerable young people in the population, or for young citizens living under authoritarian governments, the potential harms are more acute.

*

Although she lives in Europe now, Johanna comes from China, one of the most government-controlled, restrictive media environments in the world. In an illustration of the chilling effects of surveillance, she doesn't want me to use her real name, for fear that she or her family might experience consequences for speaking with me.*

Her parents ensured Johanna came to social media late. Her nature-loving father insisted she be in the 'real world', so much so that he didn't let her have a camera. He wanted her to experience life without a lens, without a frame, and perhaps with a greater freedom than she feels now. Her mother was more ambivalent: she didn't like the idea of exposing Johanna to technology but knew having it was essential for her future.

When she was fifteen, Johanna was allowed a smartphone. By that point, most of her friends had one, and her exclusion had become awkward. 'Everyone had been demanding that I use one,' Johanna says. 'There was trouble connecting with people if I didn't use WhatsApp, for example. I realised it wasn't possible to resist it any more – I was missing out on actual life interaction.'

Given how reluctant her parents had been, they were surprisingly hands-off once Johanna had the device. Understanding their daughter's personality, maybe they knew she'd be self-policing. 'They trusted me to experiment. By nature, I'm a really restricted kid,' Johanna tells me. 'And I didn't like [the phone] in the beginning.' In her innocence, she didn't know that clicking on links in junk mail or spam might take her to unexpected places. When she

---

* Again, Johanna is a pseudonym. Although from a Western perspective there did not appear to be anything particularly controversial about the views and experiences she disclosed to me, she permitted me to use her interview material only on the condition that her identity would be obscured. Her anxiety about the consequences of sharing information is an illustration of the chilling effect of surveillance, discussed previously in this book.

stumbled onto some porn sites by mistake, she immediately told her parents.

One of the first things Johanna downloaded was Instagram. When she created an account, Johanna was conscientious with her friends' privacy. 'I would always ask someone if I could tag them,' she says. 'That was odd in the eyes of my classmates. They laughed and said, *Sure. Why not? Why are you asking this?* To them, it wasn't something serious. They'd usually accept it whenever people tagged them or posted pictures about them. I didn't see it that way. I thought, it's your picture. It's your account. If I tag you, that should be with your consent.'

Johanna's instinct to seek permission from her friends makes sense: teenagers who are defeatist about their privacy being truly valued or protected by social media platforms – which Johanna was – are more likely to negotiate privacy boundaries with other people.[31] Johanna saw her friend's bemused responses as learned helplessness, borne of the belief that if everyone already knows everything, there's little point in worrying about privacy. But she also wonders whether underneath their nonchalant attitudes, there was fear. 'I think people became scared of talking to me about it because it's a scary topic for them. When I find something scary, I think we should try to stop it. But I guess for some people, it brings more helplessness to them,' she says.

Beyond privacy concerns, the way social media changed her habits in a short space of time bothered Johanna. She resented being dragged into a world from which she now feels she cannot escape. She chose to start using these platforms only because they were the price of admission for normal social participation, and she didn't like what happened to her afterwards. She found herself procrastinating, on her phone when she wanted to be doing other things. 'I got addicted. I think that's inevitable,' she says. 'Even after you log out, you're still thinking about social media. You're still thinking about the game you didn't finish.'

She's not big on games – her cousins in China are obsessed with them – but she uses social media all the time, including platforms she has serious questions about. She dislikes the way they manipulate people's behaviours, citing how adolescent girls alter their bodies in line with what they see on Instagram, some going to extremes that land them in hospital. But despite that, 'I still won't give up Instagram,' she says. 'I was forced to use WhatsApp, and now there's no way I could survive without WhatsApp. After I talk with you, I'll go back to Google to write my essays. Nobody's ready to give up on the internet.'

I ask her whether other teenagers she knows are like this – captives aware of the cage but resigned to the social contract that's been foisted on them. She admits that she and some of her close friends, graduates of international schools, might be unusual because of curricula designed to make them aware of big-picture issues. She wrote her high-school thesis on an app in China that she says has had 'a huge psychological impact on the population, like a manipulator machine for women and their consumption habits'. Her research made her more afraid for the future, about today's reality and tomorrow's scope for governments and private companies to control people's minds and lives, shaping their behaviours through apps that are socially required at least, mandated at worst.

In 2018, Johanna was watching when Mark Zuckerberg testified to the US Congress about the Cambridge Analytica data-sharing scandal, and she was wishing more of her peers would pay attention too. 'We need more voices in this area,' she says. 'I want to be a market regulator because of this.' Now twenty, a first-year university student, she feels the harms to her generation are self-evident. Where social media and teens are concerned, theoretically, things do look bad. Social comparison and the psychology of surveillance, both with well-established bad psychological outcomes, are hallmarks of our digital environment. The popular press shares Johanna's concerns, but the picture is more complex than blaring headlines might suggest. How do we regulate the bad without removing the good? And how do we figure out how bad it actually *is*?

In late September 2021, the *Wall Street Journal* released a multi-part investigatory report and podcast called *The Facebook Files*.[32] The exposé analysed a raft of leaked Facebook documents that raised serious concerns, including what sounded like damning evidence that Facebook had discovered, from its own research, that Instagram was harmful to youth, and had concealed it from the public to protect their profits. 'Facebook Knows Instagram is Toxic for Teen Girls, Company Documents Show',[33] the *Wall Street Journal*'s headline read.

The US Commerce Committee's Subcommittee on Consumer Protection, Product Safety, and Data Security reacted quickly to *The Facebook Files*, scheduling an emergency congressional hearing. The star of that show would be the unnamed whistle-blower at Facebook who had provided the materials underpinning the *WSJ*'s report. She hadn't been involved with the research on Instagram and teens, but the slide decks about the research had been available internally.[34] The whistle-blower's identity would shortly be revealed on national television.

On 3 October, Frances Haugen appeared on *60 Minutes*,[35] the primetime CBS news magazine programme in the US. Originally, she said, she'd joined the company now known as Meta because a friend had been radicalised by misinformation and conspiracy theories online, and she wanted to help create a better, less toxic, social media environment. To say that she was swiftly disillusioned would be an understatement. In a short time, she concluded that Facebook was a threat to democracy, and she was shocked by its handling of several societal harms, including political misinformation, hate speech, human trafficking and mental-health issues in teenagers.

On 4 October, in an unrelated development, Facebook and all its subsidiaries, including Messenger, Instagram and WhatsApp, went dark for about six hours. Sixty million dollars in ad revenue was wiped from the balance sheet.[36] If this was inconvenient for developed countries, it was downright destabilising for the developing world, where Facebook's 'Free Basics' programme is the main portal into the internet, providing data and phone services to about an eighth of the world's population across Asia, Africa and Latin America.[37] This series of unfortunate events sent the social media giant's stock value plummeting: on the day of the blackout it was reported that Mark Zuckerberg's wealth had dropped by $6 billion in just a few hours.[38]

On 5 October, Haugen took the stand before Congress.[39] She testified to a multiplicity of harms that she argued Facebook causes or exacerbates in society and referenced multiple vulnerable groups, but nothing seemed to grab the attention of the assembled lawmakers and the media as much as the potential threat Instagram poses to teenage girls. These threats had surfaced through Facebook's own research, which included a variety of self-report studies such as online surveys, diary studies and focus groups, in which people describe their opinions and experiences to researchers. The findings revealed that teens blamed Instagram for increases in anxiety and depression; that one in three teen girls said engaging with the platform had made their body image worse; and that among teens reporting suicidal thoughts, 13 per cent of British users and 6 per cent of Americans identified Instagram as the cause.[40] The senators' mouths were tight-lipped, their brows furrowed with concern, as they questioned her on this point. 'How many teenage girls have taken their lives because of Facebook's products?' asked Senator Ted Cruz of Texas.

'I am not aware of that research,' replied Haugen. But she highlighted a silver lining for the previous day's devastating social media outage. 'For more than five hours,' she said, 'Facebook wasn't used to deepen divides, destabilise democracies and make young girls and women feel bad about their bodies.'

In headlines across the world, the drumbeat sounded. Toxic. Harmful. Very toxic. Dangerous. Very harmful.

Dr Amy Orben isn't the type to have emotional reactions to headlines.[41] An experimental psychologist, she's won multiple awards and nominations for her research on how exposure to the digital environment affects children and teens, and whether its impacts are as intense as the popular press, parents and some of her fellow scholars believe it to be. Orben's own doctoral thesis, which won the 2019 British Psychological Society's Award for Outstanding Doctoral Research, took aim at another bogeyman much feared by concerned grown-ups: teens' tech use and its taken-for-granted negative impact on kids' well-being.[42]

In her study, Orben subjected data on 355,358 teens, twelve to eighteen years old, to rigorous statistical analysis, incorporating all manner of other variables that might also have associations with well-being, such as social and demographic factors, income, education and well-being of family members, attitudes towards school, sleep and nutrition information, and drug and alcohol use. When she ran her analysis, certain factors were indeed strongly associated with teens not doing well, such as smoking pot and experience of bullying. Other behaviours were clearly linked with teens doing better, for example, fruit and vegetable consumption, adequate sleep and eating breakfast.

The correlation of teen well-being with technology use, on the other hand, was slightly negative but minuscule compared with other activities in an adolescent's life. Time spent in the digital environment accounted for only about 0.4 per cent of the variation, and Orben calculated from this that a teen would need to spend around 11 hours 14 minutes a day on tech to experience a noticeable dip in how good they were feeling about life. Wearing glasses had a stronger negative correlation with well-being, and the link between teen well-being and tech was about the same as the connection with regularly eating potatoes.

So why the considerable panic about teens and tech? Part of it, Orben says, is our natural reaction to new things. 'Concerns about emergent technologies form a continuous cycle,' she wrote in her thesis. 'They appear when a new technology surpasses a certain popularity threshold and stop once a newer technology prompts the cycle to restart.'[43]

There's also the problem of bad science or, at least, our erroneous assumptions about how much certain kinds of research studies can prove. Most investigations into teens and technology are based on the methods Facebook used for its now notorious in-house research about Instagram: asking adolescents to respond to questions about their experiences or to opine on the *perceived* causes of their distress, a notoriously unreliable method for uncovering *actual* causes.[44] Complicating things further, today's teens have been so primed by parents, educators and general society to believe that social media is bad, they're going to say yes when asked if technology is hurting them.

For example, in 2015, a group of researchers asked over

2,000 young adolescents if they were being harmed by social media and if they were addicted to their phones.[45] Ninety-one per cent of the teens said yes, technology and social media are hurting me. But mobile ownership and social media use didn't seem to be connected to *objectively* measured indicators of well-being. The researchers collected data on the subjects' academic test scores, their feelings of belonging at school, their psychological distress levels and their physical health. There was no significant difference between the phone-owning social media users and those who went without.

Aside from the problems with self-reporting, there's a directionality problem with correlational studies like these. Does Instagram increase depression, or do depressed teens use the platform more than non-depressed ones? It's impossible to say which came first, the chicken or the egg, although there are rather more robust indications that distressed teens looking to distract or soothe themselves on Instagram tend to feel worse, not better.[46] But unless a study is exceptionally cleverly designed, ruling out other things that are going on, how can you say that a particular type or quantity of technology use is the offending factor, not some other feature of the teen's life in the modern world?

Unfortunately, most laypeople don't grasp the difference between correlation and causation. A and B may co-occur, but it's an entirely different matter to prove that A causes B, or vice versa. Researchers who should know better are prone to forgetting the distinction too when they're invested in touting the significance of their latest research paper. When non-scientist journalists in search of a scoop get hold of a juicy-sounding research result, all bets are off. Like the telephone game where a sentence gets whispered round a circle, the truth gets twisted at every translation point.

When there are news headlines about technology and mental health, Amy Orben's Twitter feed is a good first fact-checking stop.[47]

I scrolled back to late 2021 to see what she'd had to say at the time of Frances Haugen's testimony. To Orben, it was one more instance of media freaking about research that was poorly understood and likely not well-conducted in the first place, at least not well enough to draw hard-and-fast conclusions. Sure enough, it merited only a couple of article reposts from Orben, one of which was an analysis by psychologist Stuart Ritchie, who lectures on social, genetic and developmental psychiatry at King's College London. His piece appeared in the 'slow news' online publication *UnHerd*.*

'Is Instagram really bad for teenagers?' the headline read. 'The quality of the company's secret research into mental health is abysmal.'[48] And it seems that way. Facebook's research was about people's perceptions of using Instagram during 'hard life moments'. Over 22,000 people across six countries took the survey, which sounds like a big number, but because of the path through the survey, fewer than 150 teenage girls answered questions about their perceptions of body image and Instagram use. 'The headline splash of "Facebook Knows Instagram Is Toxic For Teen Girls" should, at the very least, have been: "Facebook *Thinks* Instagram is Toxic For *Some* Teen Girls" – with a subheading "On the Basis of Asking Those Teen Girls For Their Opinion",' wrote Ritchie.

After Haugen's testimony, Facebook provided its research to the public, but the release was met with suspicion: the company had added some running commentary to the slide decks. The commentary provided more context, explaining why it wasn't possible to draw any definitive conclusions from the 'hard life moments' survey.[49]

---

* 'Slow news' or 'slow journalism' bucks existing trends for fast, reactive news, focusing instead on careful and methodical investigating and information gathering, fairness, fact-checking and so forth.

People seemed to think it was too little, too late, the lady doth protest too much. 'Seeking to re-spin Instagram's toxicity for teens, Facebook publishes annotated slide decks,' TechCrunch sniped, as though it were ridiculous to suggest that you can't make real claims about anything based on a low number of survey respondents, responding to a body-image question when they'd already said they'd had 'hard life moments' in the last 30 days.[50] But a first-year psychology student could tell you there are considerable limits to what you can extrapolate from that.

If the study isn't meaningful, people said, why the secrecy? The trust in the company is now so low that *everything* looks like a smoking gun. 'It's hard to understand why, if the data is so positive, Facebook is often so reluctant to share it,' said Casey Newton, former editor of *The Verge*.[51]

But if a smoking gun causally connecting Instagram and teen girls' body image hasn't been found, that doesn't mean it isn't *there*. Some small-scale studies – not surveys, but actual experiments – have shown, for example, that manipulated photos have more of a negative effect on teen girls' body image, but mostly in girls who have an existing tendency towards more comparison.[52]

More and better science is needed, with more data to support it. Of course, companies such as Meta hold treasure troves of those data, enough to enable independent academics to discern what's going on, using the most cutting-edge, objective, machine-learning analytic methods available. With so much data, with more powerful, objective research methods and tools, data scientists could precisely, incisively identify where the risks lie for whom, and cut a more targeted path towards minimising them.

Today, globally, one in seven teenagers experiences a mental-health issue: depression, anxiety or a behavioural disorder.[53] If some of the biggest factors driving up teen stress are pressure to conform with their peers and the exploration of their identities, both of which are constantly played out on the online stage, it's ridiculous to presume that social media would have no effect on teenage mental health. Then again, the psychological effects of the online environment for that demographic could also be positive – even *net* positive – when teens use online platforms to connect with one another.

Mark Zuckerberg has argued that that's the case, in his own testimony before Congress.[54] But don't let that keep you from believing that it might be true.

The idea that the online environment is damaging for developing teens is so dominant that you don't have to be a dyed-in-the-wool doom-mongerer to unquestioningly hold what's known as the 'fragmentation hypothesis'[55] – the idea that social media use leads to an unstable sense of self through enabling and even nudging teenagers to project or play around with multiple personae. An adolescent might suddenly and frequently change their style, interests, personality or gender as influenced by communities they've encountered on social media. Able to traverse online spaces anonymously or pseudonymously, some create gaming or social media avatars to make desired characteristics manifest, perhaps freeing themselves of real-life qualities that they experience as limiting. The reverse might happen as well – in what's known as the Proteus effect,[56] a teen could identify with their created avatar so much that they start adopting the characteristics of their online avatars 'in real life'.

You can't be what you can't see, and since on the internet you can see pretty much everything, an online teen encounters innumerable identity possibilities they might wish to explore. If you believe in the fragmentation hypothesis, you'll be more likely to see this as negative, to think that a teen's chance of developing a clear, coherent sense of identity will be only undermined or challenged by so much choice, so many influences.

On the other hand, there's the self-concept unity hypothesis.[57] Self-expression and self-disclosure play core roles in identity, and during the teenage years the focus of that expression shifts from parents onto peers. Teens who can control and curate their self-image, safely explore various facets of the self and find connections with others that increase their scope for self-expression have *good* psychological and emotional outcomes.[58] For many teens, such exploration feels safer in anonymous or pseudonymous interactions, as Jessie discovered when she got support from the kind teenaged cosplayer on TikTok. Certain types of adolescents, including particularly anxious kids and younger teenage boys, benefit from being able to disclose online before moving on to offline self-disclosure. Overall, if both online and offline interaction can be helpful in social and identity development, does it really matter if online self-disclosure is less intimidating, more controllable and easier for certain adolescents at certain times than face-to-face interaction?[59]

And when a teen does want or need to talk, the chances of connecting with supportive, like-minded people online might be far richer than they are at home or at school. At the very least, social media and online communities can both augment an existing social network and provide an alternative one for an adolescent struggling socially at school, or who needs and wants wider or different circles than their offline environment can provide. During the Covid-19 pandemic, parents everywhere wrung their hands about the amount of time their kids were spending engrossed in online gaming, filming TikTok dances or endlessly exchanging memes on WhatsApp. Amy Orben penned an article in the *Guardian* pleading for calm and arguing that the screen time was keeping their kids sane by maintaining social connections.[60]

To reference Shakespeare again, technology is neither good nor bad, but connection (or lack thereof) makes it so.[61] As evidence of this, when we compare ourselves on social media with strangers with whom we're not connected, or to more distant friends with whom our connection is light, we're more anxious, lonely and insecure.[62] When we see social media posts about nice things happening for our *close* others, we tend to feel happy, perhaps basking in their reflected glory because of our relationship with them – a relationship that also gives us access to a broader, more balanced picture of both the good and bad in their lives. Passive social media use – scrolling through that Instagram or Twitter feed – is associated with anxiety, but posting, responding and building relationships on the same platforms are linked with higher self-esteem.[63]

A little while back I said that our online experiences in adolescence can have long-lasting implications for your sense of self in adulthood, which is more than just theoretical; it's been demonstrated in a longitudinal study that looked at youths between the ages of thirteen and eighteen and then again five years later.[64] The researchers recognised the double-edged-sword quality of online self-expression, its potential either to strengthen identity or to increase confusion, and decided to track how ease of online self-expression in adolescence seemed to map onto identity integration in young adulthood, measured by items from a questionnaire about 'psychosocial maturity'.

At first, the results were confusing. Expressing oneself online as a fourteen-year-old looked as though it was strongly associated with both identity integration *and* identity confusion in young adulthood. When they took a closer look, the researchers realised that an additional factor explained who went one way and who went the other. The teens who'd more successfully and deeply connected with their peers through their online self-expression were the ones with the high self-esteem and identity integration in young adulthood; the teens who'd reported *not* connecting so well, despite putting themselves out there online, suffered more identity confusion later on.

Both adolescents and adults sometimes believe that having contradictions in our characteristics, values or preferences means we're confused about identity. Many of us suffer through feeling hidebound to a particular identity, trapped by qualities we believe we must possess to be accepted, or worried that paradoxes in our personality mean there's something wrong or unresolved about us. Erikson did refer to identity clarity – referring to defined and consistent beliefs and opinions about the self – but identity integration doesn't have to mean any fixed characteristics, just a positive view of oneself, a feeling of self-worth and self-esteem.[65] Individual identities contain multitudes, including conflicting and contrasting dimensions of personality. Humans are constantly becoming, creating and recreating the self in a continuous flow of change, and Erving Goffman clearly believed in a multifaceted self, saw us as turning this facet and then another to the audience, as occasion demands.[66] The online environment has provided an environment for many of us, perhaps especially younger people, to explore, acknowledge and embrace that reality more fully.

Trying to make sense of Jessie's fascination with cosplay on TikTok, I found a book authored by three academics traversing diverse fields: creative writing, popular culture, youth fashion studies, communication and advertising.[67] 'At this point,' they say, 'it may be useful to dispense with the presumption that there is a real identity behind the constructed one, a vestigial human truth behind the mask.' The 'self' might always have been a set of performances, as Shakespeare and Goffman described, but before the internet we were able to deny that was the case, fool ourselves that we were, or that we needed to be, more fixed than we naturally are.

'The proliferation of these virtual selves', say the authors of *Planet Cosplay*, 'has helped radically overturn older, common-sense notions of who we are and what we can be.'

Perhaps there's another emotion, besides fear and anger, that lights up the brains of older generations when they look at those teenagers on their devices. Perhaps they feel something else as their adolescents rack up the screen time, accessing a vast smorgasbord of possibilities, identities and connections.

Perhaps it's envy.

Despite the limitations to what research can yet tell us in such a nascent situation, a general truth is clear: social media contains all the affordances, the whole range of positive and negative possibilities. Absolutely, there are problems with the tech pulling you in, drip-feeding you content designed to keep you there, taking the vulnerable to dangerous places without sufficient safety mechanisms. But making the situation more complex is the fact that technology's impact also depends so much on who is using it, how and why.

When headlines stoke moral panic, overstate evidence and make overblown claims about what the research shows, this is damaging too. We become incurious and dismissive about any potential benefits or protective factors that teens may derive from social-media interactions. Without good research and accurate reporting, we respond in blunt and scattershot ways, threatening to stifle freedom of thought and expression in the process. And parents' anxieties become amplified, in a way that doesn't do family relations or parenting styles any favours.

If you're the parent of an adolescent and read a scary headline about TikTok or Instagram, you become frightened. You might flip from trusting them and allowing them freedom into setting harsh limits and controls, as I did. Sometimes I skirt a difficult conversation with my own child by setting those controls via the technology, without getting her perspective or explaining mine, avoiding the inevitable negative reaction by hiding behind my phone. Social-media platforms are too new, and the research on their causal impact so much in its infancy, that we can't conclude much – yet. On the other hand, the evidence that control-orientated authoritarian parenting has negative long-term outcomes has been well established for decades.[68]

We tend to train anxious lenses on what's going on with teens, not just because of their greater volatility and vulnerability relative to adults, but because they're exiting those adults' zone of influence, which makes at least some of those grown-ups want to double down on control. Of course the stakes feel high. What happens during the adolescent period *is* consequential – in all realms of the teen's life, subject to many influences, psychological trajectories are set in motion that can extend well into adulthood.

Whatever and whomever teens interact with online and off, the most significant relationship they will discover and nurture during this time is the relationship they have with themselves. On one hand, it's beyond question that they're developing that relationship, that identity, in an unprecedented technological era. On the other, our wary, fearful reactions to new tech always trend in the same way, following the cycle of novelty, fear and acclimation that Amy Orben described.[69]

Generation Alpha is the first to be reaching maturity surrounded by, embedded in, harnessed to and connected through digital technology. But maybe those prepositions – by, in, to, through – aren't quite right. Media theorists have been arguing for years that we don't exist outside or apart from the media. We don't have 'a perceptual power and control that is independent of the media we use', as the authors of *Planet Cosplay* remark. 'Rather, media inhabits us and is coterminous with our cognitive functions.'[70]

I'll spare you the need to look up 'coterminous': it simply means having the same borders, occupying the same space, covering the same range. Never have minds and media been quite so fused. If it's always been the case that we don't inhabit some separate bubble to the informational seas in which we swim, surely this has only intensified, become more true.

But it holds across the age spectrum. Older generations tend to worry about the impact of social media on teens particularly – again, the impact of A *on* B, as though it's still possible to separate the two. Though many parents and grandparents also experience tech as an extension of themselves, the older you are the harder it probably is to grasp fully the extent of the merger, to feel in your bones the transformation that's occurred, the way elements have combined in a chemical change that can't be reversed.

In 2013, a group of scholars led by the Oxford Internet Institute philosopher Luciano Floridi released an 'Onlife Manifesto'.[71] Floridi coined the term 'onlife' to try to express the idea that 'it's no longer sensible to ask whether one may be online *or* offline'.[72] Modern teens might not know or use the word 'onlife' – to be fair, it hasn't exactly caught on generally – but they don't have to. To them, having an onlife is a taken-for-granted, socially integrated existence – the digital and physical operate in harmony, not within separate compartments or realms. A smartphone's not only a tool to be used, like a stone axe or a Phillips screwdriver; it's like a lobe of the brain, an extension of the hand, another way of sensing, feeling and knowing.

Think about the last person you heard use the terms 'in real life' or 'cyberspace', and I think you'll realise it wasn't a teen, unless they were being sardonic or mocking you. 'Cyberspace' once meant a separate plane, an anonymous zone, a faceless and disembodied realm other to your life, other to your 'real' relationships. In the 1984 book *Neuromancer*,[73] William Gibson invokes that non-reality when he calls cyberspace a consensual hallucination. Cast your mind back to the last video dialogue you had with someone you love and trust on Zoom, FaceTime, Skype or WhatsApp. You shared laughs, confidences and support. Was that a consensual hallucination? Did it feel like that to you?

Thinking about the young people who've grown up with all this, keep in mind something the Australian philosopher Patrick Stokes wrote in his book *Digital Souls*. '[Now] social media sites aren't just a place where we build a version of ourselves,' he said. 'They are . . . *part of our face*.'[74] Making hard-and-fast distinctions between concepts such as *human*, *machine* and *nature*, compartmentalising reality and virtuality, and expecting people to have a fixed, monolithic identity will never help us better understand today's adolescents.

Connected teens are hybrid in two senses of the word. First, they are interwoven entities, comprised of electricity and flesh, and second, their hyper-personal social media existence makes it easy for them to explore and embrace plural identities. A particular teen's mental health, their shorter- and longer-term well-being, might depend on whether they can confidently and coherently harmonise these identities and weave them together, or whether they continue to wander lost among them, uncertain and unsure.

6

Adulthood

'll just make sure this is off, Lorelei said. She pressed the button on her phone, and we waited for the screen to go dark.

Lorelei had an anxious mind to which she often listened, and she was prone to rumination about good things going bad, a tendency that undermined her ability to relish the joyful things when they were happening. She'd followed a tick-box approach to finding the 'right kind' of job or partner for years, but in her psychotherapy sessions we'd been revising her strategies for life, helping her be more instinctive, aware of her feelings and driven by her values.

Now, Lorelei was enjoying a happy time. She had met Leo at a casual party thrown by a friend, during a moment she wasn't thinking about meeting anyone. Now she and Leo were engaged and on the cusp of living together.

Lorelei couldn't believe her luck. Leo was handsome, funny and kind, and it seemed beyond belief that a man with such lovely qualities in his later thirties could be not only available, but as good as he seemed. She wanted to let herself relax and be happy, but she fretted that some disappointment was lurking around the corner, waiting to pull the rug from under her feet. In love but afraid, roiled with warring impulses in her gut, she wasn't sure which to follow: the conviction that Leo loved her and wanted to be with her, or the niggling fear that the situation couldn't be trusted. The uncertainty was uncomfortable, and Lorelei craved release from this agony.

She hadn't planned it. On the way to the bathroom, Leo tossed his phone onto the bed next to her. Lorelei heard the water start and the squeak of the shower cubicle door. Seeing Leo's phone shining brightly on the duvet, all its icons visible, she realised the passcode requirement hadn't yet kicked in. Lorelei did what so many of us in intimate relationships now do to try to assuage uncomfortable feelings: she checked her partner's phone.

The shower was long, giving Lorelei ample time to get into a pickle. When she didn't find anything current to trouble her, she dived down a rabbit hole. In Leo's contacts, she located a piece of information she'd never had access to before: the surname of his ex-girlfriend, Sarah. Swapping to her own phone, she tapped Sarah's name into LinkedIn. When she found her profile, Lorelei's heart sank. Leo's ex was beautiful. Worse, she was successful in professional realms that Lorelei admired.

At that moment, Lorelei's insecurities were worse than they'd ever been. How could Leo love her as much as he'd loved Sarah? She returned to Leo's phone, this time to look specifically for communications with his ex. First, she found the emails from the time they'd been in a relationship. It wasn't only *what* they said – their easy and loving intimacy triggered envy rooted in insecurity. On WhatsApp she found Sarah's perfect face, next to a message thread from a month ago. *A month ago!* Did Leo want to grab a coffee on Thursday, Sarah had wondered. *Let me check my diary*, Leo had replied.

And then – nothing. Did they ever meet? Did the coffee happen? Had Leo erased subsequent messages, afraid that Lorelei would discover something incriminating? *This isn't right*, Lorelei thought. *Something is wrong.*

As Lorelei was on the verge of looking up that article she'd seen about how iPhones can reveal everywhere someone's been, she heard the water stop. Heart hammering in her chest, she tried to replace Leo's phone exactly where he'd tossed it.

For days afterwards, she worried that something would alert Leo to what she'd done. Had she left his message thread with Sarah open? Would it appear at the top the next time he checked his WhatsApp?

Lorelei began monitoring every interaction with Leo for signs of discontent, and he had started asking her if everything was all right. The couple had never discussed boundaries around devices, but as they never used one another's computers or phones, there hadn't seemed to be a need. Still, Lorelei had to assume she had crossed a line. When it came to ethics, she and Leo were on the same page. Shared values bonded them, were one of the reasons their relationship worked – until now. Maybe they weren't ready to marry after all, she said. She told me that if she couldn't be trusted to respect his privacy, perhaps she didn't *deserve* to get married, and her feelings of shame overwhelmed her.

On the sofa across from me, Lorelei collapsed into tears. In the time it had taken her partner to shower, she had gone from a woman who was happy and excited about her future to an emotional wreck.

In the first two decades of adulthood, we're learning to navigate the complex grown-up worlds of love and work. Between the ages of twenty and forty or so, stresses around work do loom large in my practice, but the most common phrase I see in initial enquiries about therapy is *relationship problems*. If someone gives another reason for coming in, such as anxiety or depression, concerns about dating, partners or spouses almost always end up being involved as well. An unhappy relationship breeds or exacerbates other problems; a happy relationship serves as a protective factor, making us more robust in life generally.

Lorelei and Leo are a composite, a melding of several client situations I've worked with over the years. The experience of such a couple is unfortunate in a particular way, for it's a story of smoke without fire. Lorelei and Leo's relationship really was great, and Leo was as excited as Lorelei at the prospect of spending their lives together. To my knowledge, for most of the couples woven into this story, things have turned out well. But Lorelei's search, in which she found nothing definitively incriminating but enough to ignite and then stoke doubt, created a spark that could have kindled a relationship-damaging blaze.

Erik Erikson said that the primary conflict in our twenties and thirties is one of Intimacy vs Isolation,[1] as young adults work out how and whether they can love someone else and make a life with them. When you don't have intimacy in your life, you crave it, and in what might seem like a paradox, you cannot have healthy intimacy without clear boundaries. Relationships need boundaries that define and respect where one person stops and the other begins; that safeguard individual autonomy, space and privacy; and that encircle a couple, protecting the relationship from intrusions that could harm its integrity or erode the bond over time.

Our digital devices, and the ways we use them, have introduced a new primary conflict that interlocks with Intimacy vs Isolation: Boundary Clarity vs Boundary Uncertainty. The reason technology can pose such a threat to healthy relationships is because it has so influenced our creation and observation of our own and other people's boundaries. When it comes to boundaries, some people respect them and some trespass them – when a Respecter and Trespasser or two Trespassers are in relationship together, all manner of unhealthy relationship dynamics and conflict can arise.

On one hand, technology can build and maintain closeness, connecting us to people we might come to love and those we love already, including those too distant for us to see and touch 'in person'. On the other, its capability to transcend barriers is matched by its propensity to create them when lines are blurred, crossed or perhaps held too strongly. The intrusive presence of connected technologies is keenly felt in places that used to be protected inner sanctums that we shared with close others. Our devices, forever pulling our attention to the outside world and away from the here and now, continuously breach the boundaries formerly drawn around the private home, the convivial dinner table, the intimate bedroom and even the confidential therapy room.

At the start of sessions, most older psychotherapy clients put their phones out of sight. But there are some visitors to the consulting room – primarily Gen Z or Gen Alpha – who always keep their phones firmly in hand or next to them. Sometimes they clutch them when telling me about their struggles detaching from Instagram, or reducing their screen time, or maintaining attention on anything and anyone at all. I've no reason to believe that my clients' phones are any further from them when they leave the session and rejoin significant others. Sometimes I witness the invasion of devices in relationships first hand. In one memorable couples-therapy session, an impeccably and expensively suited husband fielded successive business calls before finally saying *I need to take this outside*, leaving me and his sad-eyed wife to wait in silence for his return, I helpless, she hopeless.

Phones are like third wheels in our relationships, constantly competing for our affections. Because of their nature and omnipresence, our devices play starring roles in tales of tension and uncertainty, intimacy sought or thwarted, connection or betrayal, boundaries respected or crossed. They are our allies, our enemies, our private investigators. Sometimes, they're that one frenemy who stirs up trouble by telling you something because *they'd* want to know.

Remembering all the couples I have seen for therapy over the last decade and a half, a good three-quarters were there, at least in part, because of a phone.

While this chapter won't focus on dating, it's clear that meeting partners through technology is now a norm. Stanford sociologist Michael Rosenfeld has shown through his research that modern couples are more likely to match via algorithms than through social, work or family connections and encounters. 'People trust the new dating technology more and more,' Rosenfeld says, 'and the stigma of meeting online seems to have worn off.'[2] But online or off, we spend much of our younger adulthood years looking for good relationships. We search for that *right person*, using tick lists matching our own characteristics against someone else's.

Over my years in practice, clients searching for a partner have told me about their non-negotiables, the ones they're listing on the apps. Deprived of the felt senses we have when meeting someone in the physical world, our mass migration to dating apps has forced and funnelled us even more into this criterion-based thinking. People interested in more than a brief hook-up seek concordance on personality, background, education, interests and political views, although they're still powerfully influenced to swipe right based on what they fancy in the looks department.

But which relationship-strengthening behaviours are most critical in keeping and nurturing our bonds once they are formed? The relationship research I studied in my doctorate was from a pre-digital time, and much of it was skewed from the start by existing beliefs about love. Even skilled researchers often design or interpret their investigations to fit preferred theories or hoped-for findings. Now, though, tools such as machine learning and 'computational psychiatry' can deliver more precise understandings and predictions about mental health and human behaviour, in a way that's far less influenced by human subjectivity and bias.

So, I was interested to read the results of a large machine-learning study conducted in 2020.[3] A large group of researchers gathered over forty existing psychological studies examining couples' relationship quality over time and subjected them to hundreds of hours of computational analysis. As it turns out, the kinds of individual characteristics people list on dating apps, or think they require from a partner to be happy, might matter somewhat in getting like-minded people together in the first place, but once the relationship is established, a 'good match' on personality, interests and values is relatively unimportant.

Instead, the machine-learning study showed, relationship contentment over the long haul is largely predicted by people's judgements and perceptions – how satisfied and how committed people perceived their partners to be, and how appreciative they themselves felt towards their partners, explained a whopping 45 per cent of relationship satisfaction. Conflict and sexual satisfaction also appeared in the top five important factors.

Think about a current or past relationship. How has your partner's technology use coloured your perceptions of their commitment to you, or their interest in spending time with you? Did it make you feel more or less cherished and important? Now, consider whether your own technology use made you more or less aware, appreciative and attentive towards your partner, and how they might have perceived your interest in them relative to whatever was happening on your device.

Certainly, using technology to check in and express love can increase perceptions of mutual commitment and appreciation, but it can also throw these things into question through feared or actual competition from virtual others. Sexual satisfaction could be enhanced if a couple is using that tablet on the bedside table to watch something they both enjoy, but a predilection for online porn or saucy chat that happens only away from one's partner could kill the vibe within the relationship.

Competition from virtual others is something I've seen in many couples clients. Virtual infidelities or physical ones maintained via devices are common occurrences. One husband I saw was addicted to both online porn and dating sites. He was doing his own work with another therapist, but I was seeing the couple, who came to me during one chapter of their long saga of trying to rebuild trust. This involved finding systems to keep the husband on track and accountable around his technology. When we met, his wife had his passwords, and they'd agreed she could check his phone at any time, but both that endeavour and the responsibility she was taking for his actions were exhausting her. The dynamic had changed from a relationship of equals to something more like probation officer and offender. When she went up to bed, he would often stay downstairs for a long time. He couldn't sleep, he said. Wondering what he was doing, neither could she.

Even when actual infidelity isn't the issue, though, the phone itself might be the other party, constituting a threat to the bond all by itself. While there might be some ways for a couple to use their respective phones while remaining connected with one another – playing an online game together, for example – being at home or in a restaurant with someone who's paying more attention to their phone than to you doesn't enhance your feelings of being important and appreciated. Following their gaze, you look at their phone and wonder what – or who – is so much more compelling than you.

*

In 2012, worldwide sales of smartphones exploded by 45 per cent.[4] People were trading in their more old-fashioned handsets for web-connected devices from Apple, Samsung and Huawei. That same year the word *phubbing* entered the lexicon, defined as 'the act of snubbing someone in a social setting by looking at your phone instead of paying attention'.[5] Using your phone in company was once seen as rude and offensive, a violation of social norms. The older you are, the more likely you are to still look at it that way, but we can't now say that it's not normal. Way back in 2015, 90 per cent of people had used their smartphones in their most recent social interaction.[6] And evidence suggests that we're far more likely to phub the ones we love,[7] a phenomenon that some researchers call pphubbing, with an extra 'p' for 'partner'. Its presence in relationships has been found to work against the important factors for relationship satisfaction, increasing feelings of exclusion, decreasing perceived partner responsiveness and lessening feelings of intimacy.[8]

Phone-snubbing has entered the bedroom as well – most smartphone owners charge their devices next to them while they sleep, despite many good physical, psychological and relational reasons *not* to do so, and many of us have phubbed or been phubbed in bed, with many describing how they've missed out on physical intimacy as a consequence. Sex sometimes gets short shrift when technology has reconstructed the bed as a place for work, communication with the outside world, and more technologically mediated entertainments.[9]

But whatever is happening or not happening in bed, the eyes are still the windows to the soul. Sometimes it takes a short sharp shock to make us realise how often our eye gaze leads to our phones – an incident of stepping into traffic, for example, or running head first into a lamp-post. If we don't get a handle on our devices monopolising our visual attention, though, we could find ourselves on a collision course with far more than street lighting.

From our earliest days, eye contact matters for our feelings of closeness and affiliation with others. When we're babies, reciprocal gazing with our carers or baby friends mirroring our behaviour makes us feel safe and loved and promotes trust. When researchers conduct studies with pre-verbal babies who can't reach for or grab things yet, they measure the infants' interest in or liking for things by recording what or who they are looking at, and for how long. For the rest of our lives, when we see someone we care about, we want to be seen back. A partner gazing at you makes you feel attractive and liked, and the aversion of that gaze is associated with disinterest and disengagement.

Never have our loved ones been up against so much competition for the visual attention that our brains naturally interpret in terms of love, care, self-worth, inclusion and importance. Our environment is rife with electronic rectangles that pull, capture and keep our gaze on them. The visual presence of a phone in the environment – not being used, lying dark on a piece of furniture – has been shown to have a negative impact on feelings of closeness and how much empathy we perceive our conversation partner to have.[10] In an investigation conducted in coffee shops, people watching strangers in conversation rated the dialogues where those strangers had phones in their hands as being less fulfilling.

But it isn't just that the phones we have with us all the time are so good at pulling our focus – it's how we interpret our partner's interest in the devices that matters. For example, one study showed that staring at a phone was far more harmful to a relationship than staring at a newspaper. The phones in the study had something that the static reading material didn't – they were two-way rather than one-way objects, dialogical rather than monological, portals for communication with the virtual other.[11]

For people who suffer from low self-esteem and self-worth, watching a partner's attention shift from them to potential virtual rivals is interpreted as an out-and-out threat and a driver of romantic jealousy. *Potential*, because an attempt to contact another doesn't have to happen for someone to feel diminished in the phubber's eyes, unworthy of their full attention.

Unsurprisingly, research shows that partner-phubbing is linked to more jealousy and less closeness between romantic partners, diminished conversation quality, more depression and less relationship satisfaction, trust and empathy. People who aren't in on what's happening on their partner's phone, who feel marginalised and excluded, may give up and turn to their own devices to distract them, leading to a downward spiral of reciprocal disconnection. Or the phubbee might become concerned that the partner is dissatisfied, that their commitment to the relationship may be wavering – and they might decide to act on what they're imagining to be true. When this happens, our devices and accounts have become 'technoference', a distraction in our relationships, and also the means to achieve certain ends, such as reassurance-seeking.

Not many studies have examined intimate-partner surveillance within committed, cohabiting couples like the ones I see in my practice. When you live together, it's relatively easy to gain unauthorised access to your partner's information. Computers, accounts or cloud storage might be shared among family members, with linked users operating under the tacit assumptions that members of the group won't abuse one another's trust by prying. We might give a family member biometric access to our devices to be used only in cases of emergency, trusting they'll use it only in that instance. Passing your phone over in the car so that someone else can navigate can result in journeys to other destinations. 'Shoulder surfing' – hanging about while your partner is logging into a device or account – can quickly surface the necessary passcodes, and some people are alarmingly lax about their data security, keeping passwords on sheets of paper near their computers. As levels of password protection and verification procedures become more sophisticated, hacking your partner might become more difficult, driving the more determined down more extreme avenues.

Intruding on someone's privacy isn't always done purposefully but merely because you didn't understand their privacy boundaries at first. Like many of the habits and guidelines that simply evolve within relationships without ever being spoken about, rules about access to one another's personal devices and information often aren't explicitly laid out but are learned through an experiential process of trial and error. For some, it feels self-evident that a privacy circle encompasses them and their personal devices and accounts; others assume that within intimate relationships, or even families, it's acceptable or even expected for there to be free shared access or ownership. In the early days of a relationship, someone might unwittingly break an unspoken rule they didn't realise existed. If the owner of the phone reacts with anger and upset, it's clear that the couple has failed to co-ordinate the rules they each carry about their personal information, and the discomfort that's churned up as a result is called *boundary turbulence* or *privacy turbulence*.[12]

Terms such as 'boundary turbulence' come from communications privacy management (CPM) theory, a handy system for understanding how we protect or disclose our private information across various situations, online and off. Whether it's technically or legally true, we tend to believe that our digitally held information belongs to us, that we own it, and that we have the right to decide just who gets to see or know it. We surround ourselves with many privacy circles – individual, couple, family, whole social networks – and accidental boundary trespassers experience relational consequences that usually bring them back into line.

Within close partner bonds, though, we sometimes knowingly risk boundary or privacy turbulence, breaching another's privacy barriers in the hope they won't find out. When we disregard an intimate partner's boundaries in this way, it's not always because we don't care about them or the relationship – sometimes it's because we're *so* invested that, with or without justification, we fear losing them or their hurting us. We're overwhelmed by concerns of our own, insecure or anxious itches that demand to be scratched.*

In a parallel finding with the machine-learning investigation of relationship satisfaction, one snooping study conducted with newly-wed couples indicated that people's *perceptions* mattered a lot, and those perceptions can be heavily influenced by attachment style.[13] Adult attachment styles, which may correlate with earlier patterns of interaction with parents, are characteristic ways that an individual relates to close others, including their intimate and romantic partners. The four adult attachment styles are secure, anxious (a.k.a. preoccupied), avoidant (a.k.a. dismissive) and disorganised, and the differences between them have largely to do with degree of trust and the ways people react to actual or threatened separation and physical or emotional distance from the partner.[14]

The newlywed study indicated that if you're securely attached, the levels of trust and satisfaction are high in your relationship, and you experience your partner as someone who communicates

---

* In this section, I'm speaking about loving, non-abusive, non-coercive relationships.

openly with you, you usually don't feel the need to snoop on your partner's devices or accounts.[15] If secure partners *are* worried about something, they tend towards more constructive ways of dealing with it. Willing to risk discomfort and awkwardness if it's in service of a constructive interaction, they're able to have a more open and healthy relationship as a result.

If you're anxiously attached, though, your partner telling you everything's fine and you've got nothing to worry about might not always set your mind at rest. If you also perceive that your partner doesn't disclose very much, holds their cards close to their chest, and restricts access to their devices and accounts, you're much more likely to intrude on your partner's privacy by covertly accessing their information.

In anxious-attachment situations, the need for constant verification of the integrity of the relationship might play out through technology, for example by needing the partner to be constantly and reliably in touch when they're away. But it can also take the form of constant overt and covert monitoring. For the anxiously attached who need to be sure, who feel as if they need to check, technological solutions to their anxiety are tantalising. The actual King Tantalus from Greek mythology was condemned always to be hungry, alive but starving, surrounded by laden trees whose fruit receded from his grasp and withered at his touch. The same thing happens when the anxious partner uses technological snooping to try to soothe romantic jealousy. Worried people detect and interpret information in such a way that it reinforces their worries, as was the case for Lorelei, and those with a disorganised attachment style might pursue surveillance and monitoring from the outset, already convinced that a good relationship outcome is unlikely and seeking only to confirm it. Instead, the comfort they seek always lies just beyond their grasp.

Finally, avoidantly attached adults might use technology as a shield against uncomfortable levels of intimacy, hiding behind their devices to maintain a level of closeness that feels comfortable for them. When the avoidantly attached and the anxiously attached enter into a relationship, their instinctive needs and cravings are at cross-purposes, the former drawing and defending strong emotional and technological boundaries and the latter constantly attempting to breach them.

Yet does electronically mediated snooping on our significant others actually put anxieties to rest and make us more secure? When someone's feeling uncertain about their partner's satisfaction with them and hacks into devices and accounts looking for reassurance, they might find it, but they're more likely to find ambiguous material that maintains their doubts and whets their appetite for future forays to try to settle the question.[16]

In the Lorelei-and-Leo situation, we see a collision of two broad factors: individual and technological. Her relationship wasn't actually under threat but, already anxiously attached, Lorelei made things worse instead of better with her phone snooping. A downward spiral of uncertainty and intrusiveness can be strongly detrimental to a once healthy relationship. And if that relationship is no longer healthy or has ended, the combination of intimate-partner surveillance and intimate-partner violence can be more than detrimental: it can be dangerous.

Following a break-up, one of my clients was suffering sleepless nights because of a campaign of harassment from her ex. At first, she was happy that he kept liking and commenting on her social media posts; she considered it a sign that he was taking the break-up better than she expected, that perhaps they could still be friends.

After a time, though, she started bumping into him everywhere, too frequently for mere chance. To her horror, she discovered that her phone was still sharing her location with him, something she'd allowed when they were together so that he could make sure she was on her way home safely after a night out. Worse, she uncovered evidence that he was logging into her accounts. She thought blocking him on social media and changing her passwords would sort the problem, but one evening as she was returning home from a club with friends, a strange message popped up on her phone: *AirTag found moving with you.* She found a small circular tracking device tucked into the corner of an outside pocket of her handbag.

One unfortunate outgrowth of ubiquitous digital technology is the way it enables trespassing across romantic partners' privacy boundaries through cyberstalking and creeping. When you are prey to such practices, even if you've generally been securely attached in the past, it can have serious effects for your mental health and your ability to feel safe in current and future relationships. These boundary transgressions can undermine the capacity for intimacy by compromising trust and autonomy, and when a victim must significantly curtail their online lives to prevent intrusions or safeguard themselves, they can end up emotionally and socially isolated. Things don't pan out so well for the person doing the spying, either; the more traumatised an individual is by a break-up or a rejection, the more likely they are to monitor their love interest or ex-partner online, preventing or postponing their emotional healing and recovery.[17] But what's the true scale of these problems?

While there's no strict legal definition of stalking in the UK, you know it when you see it – or when you feel it.[18] The Crown Prosecution Service describes it as a person watching, following or spying on someone in a way that distresses the victim, limits their freedom and makes them question their safety.[19] Because stalking so often involves the victim believing that they might be at risk of violence, serious injury or death – a perception that's a core element of post-traumatic stress disorder – it's little wonder that 78 per cent of people who've been stalked report symptoms of PTSD.[20]

According to the National Stalking Helpline run by the Suzy Lamplugh Trust – which was set up after estate agent Suzy Lamplugh was murdered in 1986 – 100 per cent of the stalking incidents that are reported now include at least some form of digital stalking.[21] In 2020, 84 per cent of stalking complaints were against ex-partners, with most situations representing a continuation of existing power dynamics.[22] The majority of victims said that abuse and coercive control had already been occurring in the relationship before the break-up.[23]

There are many different forms of stalking in the digital age. 'Creeping' refers to persistently or stealthily following someone online, perhaps lurking around their social media posts or repeatedly Googling them to see what comes up. Although this might feel creepy if the subject of this attention finds out about it, there's a good chance they won't. When creeping escalates enough for the other party to be aware, however, it becomes 'cyberstalking', which is more actively using technology to harass someone or hack into and monitor their devices or accounts in ways that are unwanted, obsessive and sometimes illegal. People give different reasons for why they do it, including being curious (38 per cent), suspecting a partner or ex-partner of doing something they wouldn't like (44 per cent) or avenging themselves after discovering the other person has stalked them (20 per cent). Online humiliation such as 'revenge porn' is yet another form of online abuse, the effects of which include PTSD, anxiety, depression, suicidality, isolation and other severe mental-health effects.[24, 25]

Whatever the reason, there's a lot of creeping and cyberstalking. Of over 2,000 Americans surveyed in 2020, 46 per cent of them admitted to keeping tabs on their current or ex-partner without their consent.[26] Some merely used their password knowledge, but 10 per cent had actually installed spyware to do it, enabling them to monitor the victim's text messages, phone calls, emails, location and photos. Some apps allow the installer to switch on the camera and microphone remotely, forwarding the audio and video stream to a remote device; other apps can record, stream or take snapshots of the infected device's screen. When polled about other forms of creeping and stalking, 29 per cent said they'd checked on their current or former partner's phone, 21 per cent had reviewed a partner's search history, 9 per cent had created fake profiles to keep tabs on them over social media and 8 per cent had tracked their physical activity via their phones or a health app. Around the globe, the study reported, more than one in three adults in ten countries admitted to using means such as these to check on an ex or current partner online without their consent.

If you think use of such methods is niche, think again. Cornell University found more than a million installs of creepware apps purchased from the Google Play Store over a few years.[27] In 2019, partnering with Norton, a company that sells personal cybersecurity products, a group of researchers from Cornell University's Clinic to End Tech Abuse (CETA) obtained anonymised data about billions of app installations on 50 million Android devices over several years. Constructing an algorithm called CreepRank, they ferreted out hundreds of apps that hadn't been found in research before – all manner of creepware, meaning apps whose primary purpose is to enable non-expert, non-tech-savvy users to mount interpersonal attacks.

The apps the algorithm unearthed included tools for hacking, harassment, surveillance and spoofing, which is disguising a communication as being from a known and trusted source. People who'd downloaded apps that were supposed to be for child online safety also tended to purchase apps explicitly identified as being for intimate partner surveillance. In fact, many apps marketed as tools to keep your child safe, like 'Family Locator for Android', used to have different names – in the case of that app, that former title was 'Girlfriend Cell Tracker'. Downloads of the Family Locator were highly correlated with 'Cheating Spouse' and 'Where the Hell Are You?'

Whenever the CreepRank algorithm surfaced new apps for interpersonal surveillance, the researchers followed processes to have the apps removed from the Google Play Store, but it proved a losing battle. New ones mushroomed in their place, changing up the vocabulary to obscure their true purpose.

Here's another instance, of course, where technology can be used for good. Algorithms like CreepRank could be refined, and victims notified when creepware is installed on their phones. But there's a problem. If the abuser has access to the victim's phone, then there's a reasonable likelihood that abuser might inter-cept the notification. If a way can be found to notify the victim only when the device is in their possession – using biometric checks, for example – what happens when the creepware or stalkerware installed by the coercively controlling or stalking person ceases to deliver the results, or when they receive a notification that it's been disabled? Could its removal result in an escalation into more abuse and violence?

On the Suzy Lamplugh Trust website, the introduction to a self-assessment tool poses a question that sounds as if it would be easy to answer but isn't always: are you being stalked?[28] Although you might have an inkling that something isn't right, cyberstalk-ing isn't always straightforward to detect with certainty, especially when surveillance apps are running in the background disguised as something else, such as a calculator. The website recognises that there is a conundrum inherent in potential stalking victims filling out an online questionnaire to find out if there is a problem. 'Before we start: Are you concerned about spyware or hacking?' says the initial page, suggesting, if so, that the user should get advice on virus and spyware prevention before continuing. A lurking shoulder-surfer could also see something that could cause a problem: 'If you need to leave quickly, you'll notice an "Exit site" button . . . Press this and your browser will be directed to the BBC immediately,' the page says.

The rest of the questionnaire takes you through twenty-two stalking behaviours, six of which are inherently technological: hacking, using tracking devices, using social networking, and making unwanted phone calls, emails or text messages. A further eight actions can occur with or without technology, and often happen through it: revenge porn, spying, threats, watching, third-party contact, vexatious complaints, threatening suicide and death threats.

Other behaviours are physical in and of themselves but are often amplified in power through digital technology: someone who is loitering around the victim's location, following them, visiting their home or work, or breaking into their house might have been using GPS tracking, or might be accessing the victim's home security system to spy on them, or to work out when the house is empty in order to break in or cause criminal damage.

Intimate-partner violence (IPV) is almost always paired with Intimate Partner Surveillance (IPS), and when you add in the fact that 100 per cent of reported stalking incidents in the UK involve a digital component, you start to sense the scale of the problem. Women are more likely to be stalked, and men are more likely to stalk,[29] and it has been reported in the UK that the largest IPS

category is formed of men refusing to accept the end of the relationship,[30] which deepens existing inequalities associated with the fact
that women are disproportionately affected by violence and control
in domestic partnerships.[31]

The problem with tech is that, as CETA puts it, 'The status
quo today is that technology empowers abusers, not survivors.'[32]
Victims are given advice on digital safety, told to block the stalker,
change their number and their passwords, have their devices
checked for spyware, and alter their name on or come off social
media altogether. If they stay on it, they're advised never to give
away their location. Many of these defence tactics involve survivors
curtailing or opting out of digital participation altogether, hindering
their ability to live a normal life in a world that so often demands
you be online.

To be in illegal territory under the Protection from Harassment
Act in England and Wales, all it takes is two or more incidents that
the perpetrator knows or *ought* to know would amount to stalking.[33]
The acronym FOUR is a handy tool to help you stay on the right side
of the law, or to realise when someone else is crossing that boundary: Fixated, Obsessive, Unwanted and Repeated.[34] But to have the
insight for 'ought to know', you need to see the cyberstalking as
problematic, and that's far from guaranteed. Unfortunately, cyberstalking seems to be trending towards normalisation and resigned
acceptance. The younger you are – and, perhaps, the more of an
electronic-surveillance culture you've grown up with – the more
likely you are to think that these behaviours are harmless, and the
more common it is for you to engage in them yourself. Forty-five
per cent of eighteen- to thirty-four-year-olds say they have cyber-
stalked someone, and 68 per cent say they reckon it's *okay* to do it.[35]
Furthermore, 34 per cent of Gen Z and 35 per cent of Millennials in
the US say they don't *care* if they're being stalked online by a current

or former partner, as long as it's not in person. Kevin Roundy, the Senior Technical Director of Norton Labs, terms this a 'wake-up call' and urges psychoeducation for a younger generation becoming rapidly inured to the dangers.[36]

These numbers sit uneasily alongside research by the Suzy Lamplugh Trust, showing that 91 per cent of stalking victims reported mental-health problems following their experiences, including emotional, psychological, physical and relational/social effects.[37] For the cyberstalker, at a remove from their victim, the impact on the person whose every move they're watching might not seem as salient as their own experience: their own upset, anger, or perhaps excitement or power trip. And when cyberstalkers *want* to cause their victim discomfort, someone else's pain can be an incentive, not a deterrent.

Stalking was already on the increase before the Covid-19 pandemic started, but over the lockdowns in England, incidents seemed to mushroom and intensify.[38] Although some of the higher rate might have been down to a better recognition of stalking as a problem and a crime, it also seems to be the case that stalking behaviours did increase during this period, along with all sorts of other violence and abuse. Confined to home, forced to spend more time online to try to stay in touch with the world, victims were sitting ducks at the mercy of perpetrators who had way more time on their hands.

And two years into the pandemic, the Centre for Countering Digital Hate uncovered an unsettling piece of data: a six-fold increase in UK web traffic to websites that promoted 'incel' culture, primarily an online subculture that preaches hatred of women and which has led to real-life, sometimes fatal, attacks.[39]

Dedicated incel sites and forums aren't the only places where toxic online cultures fester, of course. The CETA research group also produced an analysis of how men spoke to one another on publicly accessible online forums dedicated to intimate-partner surveillance, such as so-called 'infidelity forums'.[40] Repeatedly, they saw the pattern: someone would share an emotional personal narrative about unfaithfulness, or mysterious changes in a partner's behaviour. They would request advice and guidance. And the community would always deliver: supporting the person posting, sharing similar experiences, boasting about their own use of surveillance, advising others, and offering 1:1 tutorials and support. Spam advertisements for spyware cluttered many of the discussion forums, showing corporate interests weren't above supporting cyberstalking where there was a profit to be made. Where there were actual conversations, they made for difficult reading – the researchers found them 'violent and disturbing'.

Although they might be very different characters, common psychological threads connect the coercively controlling, angry, rejected cyberstalker and the anxiously attached or anxiously avoidantly attached snooper within an established relationship. Both may suffer low self-esteem, low self-worth, and problems connecting and trusting. Both are prone to using technology to try to gain a greater sense of power over their destinies, or to feed their cravings for an intimacy that their boundary-crossing activities will only undermine, leaving them more isolated.

We forget the link between clear boundaries and healthy intimacy at our peril; it's concerning that younger generations are becoming not only more willing to use tech to trespass across one another's boundaries but also more resigned to the inevitability of such breaches. In my past couples practice, I've helped couples calm instances of privacy turbulence and assisted them in repairing boundary ruptures. In my future work with Gen Z and Gen Alpha couples, born and raised with digital technology, perhaps I'll be persuading them of the value of having boundaries in the first place.

Young adulthood is still about discovering how to love, work and be with others as a grown person, beyond your family of origin. Intimacy vs Isolation is still a fundamental developmental crisis in this phase. But given that Boundary Clarity vs Uncertainty is such a prevalent theme reverberating for young adults in the digital age, being clear about your own personal informational boundaries – and conveying them to others – is a good way both to start and to go on, an important practice in the modern healthy relationship.

Perhaps partly because of the omnipresence of our phones, the way we now experience them as almost extensions of our bodies, we struggle to maintain this awareness or recognise the need for such conversations. In addition, anyone who's been in a relationship knows that discussions about love, intimacy and sex are not easy. When the emotional stakes are high, when the territory you're delving into provokes anxiety about rejection, it's difficult. Situations of love and commitment always have the spectre and fear of loss on the opposite side of the coin. And when discussions are hard to broach, when you always go down the same conversational patterns, perhaps ending up in conflict, assuaging your fears or anxieties through surreptitious surveillance might feel like the easier option.

But resisting that temptation is worth it. Anxious- or anxious-avoidant attachment might push you to check a phone, and avoidant attachment might drive you onto the phone to avoid uncomfortable levels of intimacy, but identifying your own and your partner's attachment style is only part of the solution: you also need to work on changing those attachment-driven knee-jerk reactions, learning to convert them to responses that are more healthy for your relationship. Intimate-partner surveillance might create the illusion of reducing anxiety and promoting trust, but it rarely has that outcome, and it gives you no opportunity to develop courage and skill in communication.

So, as a romantic partner, maybe you've been tempted to be a Trespasser, to check devices when you were feeling unsure, insecure, suspicious. Maybe you've had past experiences where trust has been damaged, and after that, it felt as if you had the right to check on things. Perhaps you arranged this between you – maybe a couples' therapist suggested it. On the other hand, maybe there wasn't anything like that, perhaps everything felt fine, but the phone was just lying there. You like to think of yourself as a Respecter, but there it was, and it was so simple to do.

How did that work out for you?

7

Middle Adulthood

In Nottinghamshire lies the ancient market town of Arnold. It boasts a library, a leisure centre and an ASDA superstore, but less prominent now are the large framework knitting factories that rose from the forests and moors with the dawn of the Machine Age and came to dominate industry in the East Midlands of England. In the eighteenth century, almost 15,000 knitting machines were whirring away in that county, providing occupation for entire families, proud of the 'fancy work' they produced with small individually operated frames.[1]

As the nineteenth century dawned and the Industrial Revolution gathered steam, machines became big enough to produce huge swathes of cloth, threatening the old ways of the textile workers. When Arnold's worsted spinning factory closed in 1810, a business that had provided thread for the local knitters and a living wage for many of Arnold's residents, it was the final straw. On 11 March 1811, tensions boiled over. As night fell, a group of disgruntled weavers conspired to do something about this technology that might eventually put them all out of work. Armed with sledgehammers, they invaded a factory floor and smashed the new frames to smithereens.

High ridges encircle the town, like the sides of a giant bowl set into the earth. With topography like that, the smashing and crashing of wholesale destruction in the factory would have sounded throughout the town, startling the populace awake. *I think it's coming from the factory*. People must have been running to their doors and windows, peering up towards the building where so many of them worked, listening to the sounds of this early instance of rage against the machine – and, perhaps, approving.[2]

The resistance spread. Soon, marauding gangs of weavers were rampaging across the country and came to be known as 'Luddites', after their supposed leader Ned Ludd. The original Luddites weren't anti-technology per se, but they wanted an occupation, a fair shot at work. Well established, skilled in their professions and proud of their products, they feared the new machines would make them redundant. In 1970, though, the *New Scientist* used the word 'Luddite' in an expanded way to refer to a general technology refusenik, someone suspicious about any novel gadget.[3] By the time Thomas Pynchon penned his famous 1984 essay 'Is it O.K. to be a Luddite?' pretty much everyone was using the term to refer to fuddy-duddies who rejected technological innovation simply due to being stuck in their ways and cynical about anything new.[4]

'Now we live, we are told, in the Computer Age,' Pynchon said in his piece. 'What is the outlook for Luddite sensibility? Will mainframes attract the same hostile attention as knitting frames once did?'

For the Luddites, the Machine Age technologies posed a danger to more than livelihoods. Working in the small-frame factories was an occupation that united families and communities, so the new knitting machines threatened social connection and cohesion as well. For many citizens of the Digital Age, new tech rattles and unsettles us in similar ways, and that may be particularly true for midlifers, who straddle a chronological digital divide.

At time of writing, midlifers straddle the line between what sociologist Marc Prensky distinguished as 'digital immigrants' and 'digital natives', groups born either side of the watershed year of 1985.[5] While pre-1985-ers might come to embrace and function well within a digital world they're essentially forced to adopt, Prensky argued that they'll be forever saddled with digital-immigrant 'accent', betraying the analogue world in which they grew up.

During Erikson's Generativity vs Stagnation crisis of middle age, making your mark on the world comes into sharp focus. Perhaps more conscious of the sands slipping through the hourglass, you may focus on both creation and procreation: putting down roots, building up a legacy, growing a family and hitting your stride at work. At this stage, Erikson reckoned, you're more likely to be settled in a home of your own, to have a career identity, and to be enjoying a greater level of financial security. If you stay stimulated and active between forty and sixty-five, doing meaningful work and connecting with others in ways you value, you're likely to be a better and happier parent, partner, employee and citizen, and you'll probably have higher relationship satisfaction, lower levels of depression and anxiety, and sharper memory and attention.[6]

Stagnation, on the other hand, is about as attractive as it sounds. Like Bill Murray's discontented weatherman in *Groundhog Day*, a man who's already feeling stuck in his boring job and then realises that he has begun literally to relive the exact same day over and over, stagnating people wonder what it's all for, what it means and whether anything matters.[7] If you're feeling ineffective, unengaged and uninvested during midlife, it's a short jump to feeling as though you're just trudging through a swamp to the end, and sometimes we fight back against that fate by having a midlife crisis, a term for which we can thank Erikson.

A midlife crisis can be kicked off by a lot of things. Your body is changing; health problems you never experienced before may start to flare. Your relationships and roles shift, and you might be fielding new challenges with the younger and older generations in your family at once. There are losses, too: you lose people you love, you lose parts of yourself and you might lose community if a work or life change upends things. Stuff happens at work, and you wonder how long you'll stay relevant, or whether you want to or can keep doing what you're doing. Conscious of ageing towards your older adulthood, it's hard to look directly at the diminishing horizon of possibility without freaking out at least a little bit.

These personal confrontations with existential realities are confusing partly because of our complicated relationship with change at this point in life. On one hand, we crave and need movement, expansion and evolution. On the other, we experience resistance: the impulse to hang on to how things are or were; the wish that nothing has to alter at all; the fear of getting old, of irrelevance, and of all the other changes that the relentless march of time brings with it. To complicate things further, with the upheavals in technology, climate and public health during the first part of the twenty-first century, it's as though the very infrastructures of human society are having their own midlife crises: work, education, the economy, communication. For someone going through their forties, fifties and early sixties at this point in history, these wider-scale disruptions either stoke excitement or feel like double trouble.

So, these days, how you're faring during the Generativity vs Stagnation phase has a lot to do with how you're navigating the new technological crisis of midlife: Embracing vs Resistance. Where you are on the spectrum from innovation-loving Technophile to wary Neo-Luddite depends at any age on a collection of factors: your personality; your attitudes and beliefs about tech; the specifics of your circumstances at work and at home; how powerful or powerless, privileged or deprived, connected or isolated you are; and the actions you actually take. These things matter not only for work, but for your experience of staying connected with others in general – which is where I'll start.

If you've ever taken a personality test, the chances are it was based on the 'Big Five' personality dimensions identified by American psychologist Lewis Goldberg in the 1980s, summed up by the acronym OCEAN: Openness to Experience, Conscientiousness, Extraversion, Agreeableness and Neuroticism.*

Personality was once thought about as being consistent across the adult lifespan, with our traits hardening to concrete somewhere around our thirties. More recent research tells a different story. Maximally malleable as children, we tend to become increasingly stable as we move through our twenties and thirties. A strong core of personality grounds us, but just as the built-in flexibility of trees and tall buildings keeps them from toppling when the wind blows, the capacity to bend and change helps us roll with life's punches.

During the forties and fifties, though, many of your personality dimensions reach 'peak stability', particularly Openness to Experience.[8] If you come into your forties high on the Openness to Experience personality dimension, your usual curiosity, willingness to embrace novelty, and creative, proactive attitude to change will probably remain relatively constant for the next couple of decades. On the other hand, if you enter your life's midpoint as someone who's typically cautious about new things, possessed of a traditional mindset and more comfortable with the familiar,

---

* A newer six-factor model, HEXACO, has been developed since this time. However, the main change (the addition of the Honesty–Humility factor) isn't as important for consideration of openness to technology.

it's those personality features that become yet more fixed in your middle years.

When the Covid-19 pandemic's lockdowns began, digital technologies became a lifeline to the outside world for many of us – 90 per cent of Americans in a Pew Research Center survey said that the internet had been very important for them personally during the pandemic, although they worried too: fretting about their increased time online (33 per cent), sometimes wearying of video communication (40 per cent) and worrying about the costs of home broadband (26 per cent).[9] For those on the connected side of the digital divide, everything happened onscreen: gathering with friends, family and community; working; attending live entertainment, exercise classes, religious services and funerals; and consulting the doctor.[10] Journalists and pundits – prone to speaking about technology as though it has a monolithic, predictable impact – wrung their hands about what this surfeit of online connection would *do to* us.

As restrictions and lockdowns continued, I was struck by the huge variety in people's responses to the pandemic. Some differences were related to access (or not) to devices and fast broadband connections, and to varying levels of digital knowledge and competence. People had diverse home situations: living alone, locked down with close and loving family or friends, or trapped in unhappy or even abusive households. Physical amenities, including enough space, privacy and access to outdoor spaces, mattered too. But I knew of people in each one of these circumstances who coped well or badly, better or worse, because of the *internal* factors mediating their experience.

During the most difficult days of the pandemic, Big Five personality factors played a significant role in whether individuals embraced or rejected technology as a protective factor against loneliness, isolation, boredom and other lockdown depredations.[11] Being high on both Extraversion and Neuroticism, I would have expected to be extraordinarily lonely and riddled with anxiety during lockdowns. But my high Openness to Experience qualities protected me, pushing me to create and participate in stimulating, novel and diverse activities online and to focus my technologically mediated interactions on quality time with close others.[12] By contrast, a relative in older midlife, just as high as me on Extraversion but quite low on Openness to Experience, was averse to any departure from her routine, even though the restrictions had already changed it whether she liked it or not. Even though it meant being lonely and bored, she refused to join the virtual coffees and online exercise classes to which she was invited, saying she preferred to wait until things returned to her preferred normal.

Plenty of my fellow psychology and therapy professionals – many of them midlifers and seasoned professionals like me – wanted to wait for the return of 'normal' as well, and it wasn't just personality factors that influenced them: it was attitudes and beliefs, influenced by learning and experience. Even those more recently trained had been taught that therapy should happen face to face, and that certain conditions – including being together in the same physical space and holding absolute confidentiality, which many of them were unsure about online – were optimal and even necessary for therapy. These conditions are known as the 'therapeutic frame', and when I was learning my trade in the 1990s, I was taught that breaking the frame would disrupt the deep relational connection that is the bedrock of both therapy and other fulfilling, nurturing social connections.[13]

If Sigmund Freud had enjoyed high-speed Wi-Fi in early twentieth-century Vienna, maybe the founder of talking therapy would have formed different ideas about the conditions that were needed for minds to meet and psyches to heal. But that wasn't the case, and traditional assumptions about what's required for true human-to-human connection and therapeutic effectiveness are hard to shake, despite a plethora of research studies showing that online interaction can facilitate meaningful emotional and psychological contact. In one carefully designed lab experiment, for example, research participants met with a stranger in one of two situations: an offline 'face to face' encounter or an online conversation. The stranger, who was really part of the research team, made a deeply personal revelation during the interchange. In both conditions, the disclosure was made in exactly the same way, using the identical words. In one condition, though, participants reported more intimacy and feeling closer to the stranger who had been so vulnerable with them. You guessed it – or did you? The *online* participants felt more connected with the person who'd confided in them.[14]

When I interviewed a group of psychologists in 2012 about their attitudes towards digital technologies, though, they hadn't read that research. At that time, there were no restrictions on working in the physical office, and the professionals I talked to had little to no experience of technologically mediated work. Despite their lack of exposure to it, participants all said they believed tech interfered with 'real' relating. I developed a questionnaire from their responses, the Digital Age Technologies Attitude Scale (DATAS), to capture attitudes, behaviours and beliefs about tech on a scale ranging from total resistance to enthusiastic embracing.[15]

Before doing my research, I'd assumed that the older a practitioner was, the more resistance to technology they would have. Interestingly, this didn't seem to make a difference. In fact, this is one reason why Prensky's categories of pre-1985 'digital immigrants' and post-1985 'digital natives' have fallen out of favour. Uptake of technology, and attitudes and beliefs about it, vary within every age group and are better predictors than age for whether we embrace technological progress and whether we respond to challenge. If we must divide people up, it's been argued, sorting them into 'digital residents' and 'digital visitors' is probably a better way of doing it.[16]

Part of the reason the psychologists' aversion and cynicism bothered me is that what *they* believe – with or without solid and up-to-date research-based evidence – has a powerful influence on what the *rest* of us believe about technology and connection. When a blogger or journalist is writing an article about technology's psychological and emotional impact on us, what do they do? They ask an expert in mental health, a psychologist or therapist, who might themselves be low on something called 'PROI'.

PROI stands for the *perceived reality of online interactions*,[17] which is essentially what I was trying to measure in developing my DATAS scale. High-PROI people see online contact as potentially just as fulfilling as physical, face-to-face conversation, and focus not on *whether* technology can provide them with what they want from an exchange, but *how* they can use the technology to get their needs met. In contrast, low-PROI people assume online contact will be inherently less real, good and valid compared with physically co-present interaction, no matter what, and that it won't give them what they need even if the technology quality is great. Low-PROI-ers are a major genus in the neo-Luddite family of technology refuseniks. You all know someone – and maybe it's you – who says they can't *really* communicate if they're not in the same room with someone, or that it's not an *actual* conversation if it takes place through a screen, or that they can't discuss important or sensitive things unless it's *in person*. Unfortunately, that anxious avoidance can keep them from getting what they need.

For example, my therapy client Martin was a traditionalist, plagued with painful nostalgia, a yearning for what he saw as simpler times.* On the long list of things he mistrusted, digital technology ranked high. The mobile he carried was unconnected to the internet, the type of 'dumbphone' I once saw behind glass at the Design Museum in London. When not making a call, he kept it switched off. He did, however, have a home computer and became a late adopter of Twitter, warming to it quickly. In fact, he'd heated up so rapidly that in short order he was permanently suspended from the platform for infractions he never disclosed to me. The ban only deepened his conviction that technology and the companies controlling it were out to get him.

Martin and I met infrequently in the years prior to the pandemic, but he trusted and relied on me. In our sessions, he derived a lot of comfort from venting his anxieties and frustrations, but he was resistant to my attempts to help him more deeply by addressing his fixed thoughts and beliefs about the modern world and his powerlessness within it. In Prensky's terminology, both Martin and I were digital immigrants, but while he was a digital visitor, I was a digital resident.

When the lockdowns happened and I moved my clinics onto Zoom, I anticipated Martin might be worried, but he valued the work so much that I assumed he'd come round. I sent him my 'Guidance for Remote Psychotherapy' document, pointing clients to the research about its effectiveness, telling them about my privacy and safety policies, and giving them tips about how to get the most out of it, how to help forge as natural a connection between us as possible.

Martin reacted and responded with distress. First, he was astonished that I'd bought into what he was convinced was fake news

---

* Martin is a pseudonym, and other aspects of the story have been changed to preserve his anonymity.

about the virus. Second, although he couldn't express exactly why the state might be interested in the ups and downs of a private citizen's emotional life, he reckoned that the government must possess the key to de-encrypt online conversations, and that they wouldn't be shy about using it. Overall, though, he simply didn't believe the therapeutic process was possible online. *A conversation through screens isn't a real conversation*, he said scathingly. He said he would wait.

But even though he desperately wanted sessions to continue, his resistance proved to be the end of our work together. Like so many of my colleagues, including those who were initially sceptical, I never returned to the physical office. I still get emails from him occasionally, enquiring whether that's changed. But the office rental fees aren't worth it when demand remains so low – the percentage of mental-health appointments that take place online has continued to rise, even after lockdowns ceased. Like one headline in *Psychology Today* says, 'The Data Are In – Telehealth Is Here to Stay'.[18]

I and most of my clients were able and willing to embrace the novelty in service of being able to continue to meet, but Martin resisted. His attitudes were too low-PROI for him to give technology a chance. I often wonder, though, what might have happened had he been younger, before his characteristics and beliefs had hardened in midlife. Had he been younger and more flexible, might I have been able to persuade him to give it a chance, and might the experience have convinced him to re-evaluate not just his fixed ideas about technology, but much else besides?

As entrenched as I knew Martin's views were, I really wanted to get both him and my other more tech-averse clients to move the needle on their Openness to Experience dimensions of personality, because for people who were both open *and* high-PROI, the cloud of lockdown had some surprising silver linings. Some even emerged more socially and creatively fulfilled than they'd felt before, or made decisions about life or work that are serving them well.

Having not been able to convince Martin, I tried a different tack to invite others to challenge fixed ideas and fears about technology. In early 2020, I wrote an article summarising why technology *could* help us meet a lot of our natural social needs.[19] Used deliberately and thoughtfully, I said, technology could help prevent social isolation and the loneliness associated with it.

I also pointed to evidence that online interactions could give our bodies some of the feel-good social and bonding hormones so important for humans, which felt important to emphasise given widespread concerns about 'touch starvation'.[20] Professor Paul Zak, a neuro-economist who's spent years studying the neurotransmitters of connection, has found in multiple studies that we don't need to be physically face to face with someone, or hugging or touching them, to get a release of oxytocin. 'Your brain interprets tweeting as if you were directly interacting with people you cared about or had empathy for,' Zak writes. 'E-connection is processed in the brain like an in-person connection.'[21]

And if a burst of feel-good hormones can happen because of pro-social, good-feelings interaction on Twitter, Instagram or any other platform as Zak's research demonstrates, imagine the potential for online interactions with the people that you know, love and trust.[22] When a hug or a touch is impossible, the key to maximising contact-free oxytocin boosts is the emotional content and quality of the conversation.

Conversations that are merely transactional, the kind you can have with a call-centre representative, a chatbot or a search engine that delivers the information you asked for, are unlikely to provoke a surge of bonding hormones. Nor will interactions such as Zoom meetings where you're pitching your product or perspective or service, attending a committee, or trying to get a job done. But there are also transformational dialogues, full of care, deep listening, sharing, exploration and generativity, and those are a different animal altogether. During the pandemic, people who had access to the tech and deliberately used it to feed their brain and body with the best-quality social hormones they could at the time did better, and I felt it was important for people to realise that there were many possibilities for technology to help, that it wasn't all bad.

Not everyone appreciated my arguments. Although I was explicit in saying that interaction using tech was not a like-for-like substitute for physical contact, some commentators simply weren't buying that communicating with others using devices held positive potential for meeting our social and physiological needs. At first, rather prejudicially, I noticed that the most vociferous, negative and dismissive commentators had profile pictures placing them solidly in the midlife age bracket, but then I reconsidered. There are plenty of potential reasons for such a correlation that aren't about age.

As those of us with unfettered access to the internet sometimes forget, connectivity and device ownership isn't guaranteed or affordable for all. Digital poverty is one of the many existing inequalities whose effects have been worsened by the pandemic. Whether because of geography, financial circumstances, or lack of skill or education, over 6 per cent of UK households have no access to the internet.[23] In life-stage terms, older adults are more likely than midlifers to be without access, but plenty of people at midlife lack the funds for internet bills,[24] and the poorest were twice as likely as the richest to feel isolated and lonely in lockdown, partly due to lesser access to digital devices and connection.[25]

Physically and socially isolated midlifers, living alone or in loveless households during the lockdowns and without good technological bridges to the outside, would have struggled without connection in more ways than one for an extended period – a powerful argument for addressing ongoing digital division and digital poverty. When the access *is* there, though, we *can* give and receive love down the broadband line, to such an extent that our bodies might naturally respond as if the person were actually there. Especially in times such as the recent global-scale midlife crisis, when physical meeting simply wasn't possible, that's important to remember. Digital-immigrant, digital-visitor, low-PROI, low-Openness people of all ages, take note.

Most middle-aged people today were probably socialised in their younger years to aim for 'stable' jobs or careers as their parents and grandparents had. When Erikson initially developed his stages, a 'job for life' was commonplace, so we can understand why he assumed that greater stability in midlife was likely.[26] No one prepared today's midlifers for a 'gig economy' – a phrase coined during the 2009 financial crisis – or foresaw the widespread remote and hybrid working they would encounter down the road.[27] Their secondary-school career counsellors didn't talk to them about side hustles or becoming multi-hyphenates[28] – a term that's now spread outside its Hollywood origins (actor-singer-dancer) into other sectors, including the world of online creatives, content producers and podcaster-vlogger-Instagrammers.[29] They weren't warned to expect future technological innovations so dizzying that, in the prime of their working years, many of them would find themselves upskilling, pivoting and reinventing themselves again and again in a climate of near-continuous occupational disruption and change.

The pandemic accelerated the demise of 'jobs for life' by propelling rapid upswings in automation, AI, remote work and e-commerce.[30] According to a McKinsey Global Institute report on the post-pandemic future of work, up to 25 per cent more workers than previously estimated will probably need to switch occupations imminently as a result.[31] When Erikson spoke of the generativity in midlife, he was certainly assuming a more secure and less volatile foundation.

I had my own midlife professional reckoning in my late forties, before the widespread upheavals of the early 2020s. I was feeling frustrated with my job. I had to hit burnout to acknowledge it, but eventually I decided it was possible to take a leap. Over years of work, I had accumulated not just experience but wisdom. I had far more self-knowledge than in my twenties and thirties. I accepted more fully, as I approached my fiftieth year, that my life wasn't infinite. If this was a crisis, it was one with a good result: it pushed me to ask stagnation-busting questions about why I was doing what I was doing, whether I might do something else and how I might thrive.

When I quit my university job, I shifted more time to my psychotherapy practice. I saw people from various backgrounds doing all manner of work, but most of them occupied the same midlife age bracket as me. They came in with anxiety, depression, self-doubt and relationship issues, but interlinked with all of these were issues with employment: stress and burnout, overwork, bad management, and just plain not liking or caring about what they were doing but feeling stuck. Meanwhile, multiple friends were going through the same things. Even when there were numerous viable alternative roles to explore, they'd find reasons to stay in their suffering. These were smart people, but to paraphrase the title of a book about this phenomenon, if they were so smart, why weren't they happy?[32]

You might have noticed that sometimes you make choices based on what *should* make you happy, not what *does* make you happy. But, as you also might have noticed, identifying this problem often isn't enough to change it. Many hard-working midlifers postpone happiness, adventure or taking chances instead of prioritising these things in the now. When you have dependants, a home and/or other commitments and obligations, you can easily become the ant of Aesop's fable, toiling to lay away stores for the winter of life because that feels safe and secure. You can take your foot off the gas later, have fun in retirement, sleep when you're dead. And so, the weeks slip into months and roll into years, life churning you through the machine like the factory cogs that ensnare Charlie Chaplin in the famous scene from *Modern Times*.[33]

When Covid-19 skewed the working world off its axis, midlifers who'd long had too little time for contemplation and too little courage to alter the course of their working lives proactively found it far easier to change when change was suddenly imposed on them. All bets were off, rules and assumptions were broken, and the usual expectations were suspended. Overnight, people's fears dissipated about how it would look on LinkedIn if they quit a hated post 'too soon' or made a weird, radical move of job or career out of their usual sector. Many were finally emboldened to follow through on what countless happiness-research studies have long pointed to as a superior strategy for contentment: like a flower rotating in the direction of the sun, they began turning away from extrinsic, material markers of 'success', or fixed ideas about what they had to do or should do.[34] They opened up to the idea of what a more workable way of working might look like for them. What could they do? How could they thrive? Perhaps it had been a long while since they'd asked themselves these questions.

Unprecedented numbers of people changed, pivoted sector or quit their jobs altogether during the Covid-19 pandemic. Pundits dubbed it the Great Resignation.[35] During three months in 2020, over one million workers in Britain moved jobs, and about 400,000 left, the largest resignation spike in UK records.[36] In 2021, an unprecedented 47 million Americans left their jobs.[37] The exodus has showed few signs of stopping. In March 2022, the consultancy firm PwC undertook a Global Workforce Hopes and Fears Survey, tapping the emotional landscape of workers in forty-four countries and territories. They found that 20 per cent of workers were planning to leave their employment sometime that year.[38]

But why were so many of us doing it? What were we after? Asked that question, most people (71 per cent) cited pay. Many aspects of life dear to midlifers, however, also made significant appearances: desire for fulfilment (69 per cent), the chance to be more truly themselves (66 per cent), and to be creative and innovative in their jobs (60 per cent).[39]

What had whetted our collective appetite for a fresh start? Was it just the result of the sudden, enforced pause? Was it fuelled by thinking that there might be more to life, a life we had realised anew was fragile and transient: nature, leisure, family? Many of us decided we wanted more of those things in the balance, and some of us began to understand it wasn't necessary to *leave* work to get bigger slices of the quality-of-life pie.

For ages it had been technologically possible to perform many kinds of jobs on a remote or hybrid basis. Old work patterns and habits die hard, though – like Prensky's categories of people, some jobs and work cultures are digitally native, some digitally immigrant and some are entirely offline. The prevailing assumptions of digital-immigrant jobs had perhaps prevented us from availing ourselves sooner and en masse of the fuller extent of remote-working possibilities. But then things changed.

In his 1959 book *The Landmarks of Tomorrow*, the father of modern business management, Peter Drucker, coined the term 'knowledge worker', a kind of labourer who generates value through what they know. High-productivity, high-creativity knowledge workers would be, Drucker predicted, the most valuable assets in the coming twenty-first century.[40]

At the dawn of the second pandemic year, the global economy was powered by over 1 billion of these knowledge workers,[41] and it started to occur to them that while they might *have* to work remotely at that moment, eventually they could carry on because they *wanted* to. For the majority of people who'd had a taste of working flexibly, the answer was an emphatic yes. Many respectfully declined or fought back when invited or encouraged to return to the office full time. In a 2022 McKinsey survey of 25,000 Americans, it was found that 58 per cent of them could work from home at least one day a week, 35 per cent were able to work from home full time and 87 per cent would take the chance to work flexibly if offered it.[42] This went for all sorts of jobs: 'white-collar' work, knowledge work and certain 'blue-collar' jobs that, once upon a time, would have been perceived to require work on site. Peter Drucker would have approved of this. Central to his management theory was the idea that nurturing and empowering individual workers' choices and needs were far more important than rigid rules and working structures, and even profits.[43]

The people most fully embracing home or hybrid working seem to be midlife knowledge workers – or, as the *Telegraph* put it in May 2022, *Still working from home? You're probably middle-aged and wealthy*.[44] Despite the McKinsey finding about the variety of people who can or want to work from home, people in lower-earning brackets – workers in the service and hospitality industries, manual labourers and many in the creative arts – often can't choose flexi-working at all.

Yet half the jobs that exist today will probably have been transformed by automation and artificial intelligence by 2030, in ways that are difficult to predict now.[45] Instead of being replaced by computers, most of us will need to work alongside rapidly evolving machines. When the coming wave of quantum computing bursts the current confines of binary code, enabling extraordinarily fast computation, this evolution is likely to step up further. Jobs that would have sounded like gibberish to my ten-year-old self are touted as the careers of the present and future: cyber-security expert, software developer, UX designer, digital data analyst, e-sports coach, smart-meter fitter.

If the original Luddites were discomfited by the relatively stately pace of knitting-frame innovations, today's more traditionally minded Neo-Luddites are discombobulated by far more mind-boggling technological leaps, and perhaps they're not always overreacting. The speed and scale of change may mean that later midlifers, in particular, *should* be more concerned about redundancy or unemployment. As open as you are and as curious about technology as you might be, changing things up isn't always straightforward, practical or affordable when you're closer to retirement. My partner, having worked the entirety of his life on petrol-drinking internal combustion engines, sees an imminent career dilemma in the news stories about cutbacks in the UK automotive industry.[46] He wants to work for years yet and would feel unfulfilled otherwise, but if the price of continued employment were retraining in electric vehicles, would he be willing and able to pay it?

For a long while, I wasn't worried for myself. I do psychotherapy and write, and I assumed that these and related professions – vocational, creative and human-to-human jobs like mine – wouldn't change too much with technology. Maybe you think your job is similarly immune to technological advances, that they wouldn't prevent your thriving in work because there's just not that much technology involved. Maybe we both have more to be worried about than we think.

In the United States, the average age for a hospice chaplain is fifty-one, and it's generally similar in other countries, for understandable reasons.[47] A job of that kind requires maturity, life experience, great sensitivity and intense empathy. Sitting with a dying person, you have conversations of the deepest import; you carry the responsibility and honour of being alongside someone while they are poised on the cusp between life and death, while they wrestle with some of the most profound existential fears and questions imaginable. To sit with someone in such moments is to be part of a sacred interaction in a sacred space. Of course, it's a livelihood, but to term it 'just' a job feels inaccurate, a betrayal of that more important layer. You cannot help but understand it as a *calling*.

In August 2020, with Covid rampaging and death and isolation everywhere, a group of hospice chaplains from Minnesota received an email from their employer.[48] The health company had already been deploying productivity tracking to an extent, but the perceived inequities in people's workloads was problematic, and pandemic life had made it harder for management to keep tabs on what their employees were doing throughout the day. The bosses would now be subjecting their hospice workers to far stricter protocols than they'd used in the past.

As reported by the *New York Times*, the management of this company took the complex services their hospice workers performed for dying people and their families, broke them down into specific functions, and assigned each function a value.[49] A visit to a dying individual would garner you a point, while a phone call to provide solace to a grieving family was worth a quarter of a point. At the start of each working day, a worker would predict how many productivity points they would accrue by evening. Each evening, the productivity software tabulated how close they'd come to their goal.

But, as the article put it, 'dying defied planning', and sometimes the productivity gaps were great indeed.[50] Whatever points might be sacrificed, however much it might dock their pay or prevent a raise, the hospice chaplains' consciences often constrained them from prematurely leaving the sides of dying people who needed them. Some decided that this tracking, these metrics, were preventing them from following their calling. I can only imagine the spiritual struggles they went through, the dark nights of the soul they endured. In the end, they did the only thing they could. They quit. If this was how their profession was evolving, they couldn't evolve with it.

I wonder where they went, whether they changed their role entirely, where they could go to escape this bean-counting, or at least counting of the wrong beans. In America today, eight out of ten private companies use technology to capture productivity metrics for their workers, frequently in real time.[51]

For warehouse workers and delivery drivers, the metrics are more objective: number of boxes packed, number of deliveries completed. Workers such as these have been long accustomed to their worth being measured by such indices. Such ideas weren't born yesterday. In my own late 1970s childhood, adverts for Domino's Pizza proclaimed, 'Delivery in 30 minutes, or it's free!' – a promise to the customer that doubled as a productivity and efficiency metric for its employees.[52] After the deaths for Domino's delivery drivers from dangerous driving climbed so high that their mortality rate reached the level of miners and construction workers, Domino's quietly dropped the guarantee.

But Jodi Kantor, one of the journalists reporting on the rise of employer surveillance in such unexpected professions as hospice work and psychotherapy, describes how a new awareness is dawning on people in what we used to call 'white-collar jobs', graduates of colleges and universities, with advanced and specialist degrees. On *The Daily* podcast, she tells host Michael Barbaro about how her sources were experiencing the kind of frustration, lack of power and lack of control that lower-paid, less-skilled workers in companies such as Amazon have bemoaned for years.

'They're talking about all the feelings that monitoring provokes,' Kantor says. 'To what degree is work just transactional at the core? . . . Am I just a bunch of clicks? Are you?'[53]

When Erikson talked about midlife Generativity or Stagnation, he was thinking about identity and the experience and desires of the self. When I think about evolving or resisting technology in the middle years, I'm thinking about individuals and their emotions. Psychologists like us presume work is key for well-being, that well-being *matters* and that we should care about it. But maybe that's too idealistic a view of what's going on in the bigger picture, particularly in the current era. What's the more cynical view? As Nikolas Rose says in *Governing the Soul: Of the Private Self,* 'The worker is no more than a factor of production, just one factor among many caught up in a process whose sole rationale is profit.'[54]

Harsh, but perhaps he's right. In capitalism, at least, the bosses have one fundamental aim. Extract as much as you can from the workers, while minimising their resistance to the central goals of the corporate machine: efficiency and productivity. If this is the reality, it doesn't mesh with fantasies I indulge in myself, and it doesn't fit with the conversations I have with the people I treat or coach. We speak earnestly about how work needs to *mean* something, how you *deserve* to thrive, to prosper and flourish on your own terms, in accordance with your own values. Surely, especially at midlife when you're building your legacy and solidifying your place in the world, you need to be *happy* at work. Otherwise, what's the point of it all? Work doesn't have to be ugly, utilitarian, non-human. One of my close affiliations is with a collective that aims to 'make business more beautiful'.[55]

Laid out like that, it seems ridiculously privileged. Likely, the warehouse packers of today are no more romantic or deceived about the nature of their relationship with their employer than were the frame knitters on the floors of nineteenth-century factories. Nikolas Rose suspects that it's people a bit like me, or at least adjacent to me – leadership coaches, occupational psychologists and the like – who over the years managed to cast 'a cloak of legitimacy over the fundamentally exploitative nature of employment'.[56]

But then Covid-19 sent workers home to work on their own recognisance, and many employees began overhauling their perspectives and practices around work. This was a problem for employers who felt that keeping close tabs on their workers was a key element of protecting their profits. At this physical and psychological remove, employees were suddenly exercising a far greater degree of decisional control over their working days. They were giving more thought to notions of happiness, of work–life balance, of the work getting done whichever way they chose. The sins of some lightly supervised and minimally accountable employees do make for rather sobering reading. Released from direct oversight, they spent their paid time playing video games, watching pornography, using bots to mimic typing and 'mouse jigglers' to look engaged, moonlighting for other companies, and subcontracting their work out to lower-paid people.[57]

In the minds of managers, this was rather too much indulgence for comfort, and you can see the problem. How can you pick out the rotten apples if you don't go through the barrel? Because it provides an affordable and efficient answer to that dilemma, it's little wonder that employee surveillance is now becoming standard, cast as a business necessity in a remote-working world. As is so often the case, however, the strategies we use to try to solve a problem often end up with us in a place worse than where we started. Being monitored is, essentially, to not be trusted, and not to be trusted is not to be happy.

In a study reported in *Harvard Business Review*, researchers tried to determine whether surveillance was delivering the improvements in performance and accountability that employers want.[58] First, they asked employees about their experience, and they reported being more likely to ignore instructions, damage property at work, pilfer office equipment, sneak unauthorised breaks and work slowly on purpose when they were being watched.

To see if surveillance was actually *causing* such behaviours, the researchers designed an experiment. Two hundred employees were asked to complete a task, with half informed that their work would be monitored. Given an opportunity to cheat, the employees in the surveilled condition were *more* likely to take it. Why? 'Our studies showed that monitoring employees causes them to subconsciously feel that they are less responsible for their own conduct, making them more likely to act immorally,' the researchers concluded.[59] Lack of control and agency can be extremely dehumanising, anxiety-provoking, depression-inducing experiences. We find disempowerment so troubling that we find creative ways to get around it – metaphorically, we outwit and even sabotage the machine.

Besides, there's another problem. Surveillance works only if the work it's surveilling is performed on a digital device. If there aren't keystrokes to count, mouse movements to tabulate and on-screen presence or engagement to measure – if it doesn't happen on a connected device – *it doesn't count as work*.

In the investigative feature article about employee surveillance in the *New York Times*, there were depressing tales of the therapist whose work with drug addicts wasn't acknowledged because she was having face-to-face conversations with them.[60] Then there was the writer whose pay was docked because sometimes he used paper, for the legion benefits of handwriting for thinking, creating, remembering and developing ideas don't matter to the productivity-tracking machine. And, of course, there were the hospice chaplains whose noble and even sacred work with the dying doesn't obey the rules of the factory floor.

The deal some of you will end up making with your employer looks like this. You can carry on working from home, but in return you will be electronically micromanaged, held accountable for every delayed keystroke, for every camera capture in which you are absent from your desk. Whatever control and agency you thought you were gaining by 'flexible' working erodes before your eyes, and suddenly you can see the deeper bargain that you've struck, which you never saw with such clarity before. *This isn't about you. We don't care about your calling, unless having one helps you feed the machine.* Yes, your working life is indeed being judged based on a bunch of clicks.

Is being resigned to that fact how 'progress' manifests in the modern world? Or does accepting such things lead to the worst kind of stagnation imaginable? In some instances, when what's going on with tech doesn't feel okay, *should* we resist? As the quotation attributed to Krishnamurti says, 'It is no measure of health to be well adjusted to a profoundly sick society.'[61] One wonders whether the Great Resignation has partly to do with a refusal to adjust to something that feels toxic.

My father is glad to have practised his vocation in the 'good old days'. As he sloped through midlife towards the end of his career as a doctor, he felt as though the rot was setting in: the all-powerful insurance companies scrutinising claims, always with an eye to denial, the endless assigning of codes to procedures. As the cold transactionalism of modern medicine descended, it interfered with the kind of care he most wanted to give to patients. His retirement, he says, couldn't have come at a better time.[62] If things were going to keep evolving in that direction, if that's what the practice of medicine was going to become, he didn't want to embrace it.

I can imagine that, if I were a bit older myself, oversight of the kind described above would push me to kick-start my own retirement a few years early. I once thought psychotherapy was immune to automation. Technology always served my writing rather than limiting it, and only in the last year have I begun worrying about being replaced by machines. No employer surveils me or counts my keystrokes, although sometimes – like some workers who report that they like being monitored – I wonder how being overseen might help me, minimising the siren call of electronic distractions and the attentional weakness brought on by the chronic distractions of technology. It's hard to imagine how technology could ever take away my professions, or my love of them.

But I bet the hospice workers also assumed their vocation was inviolate, and I now read that virtual-reality therapies are developing fast for conditions such as stress, anxiety and trauma. Human therapists may one day take a back seat to the simulation, simply standing by to provide quality control or the occasional intervention.[63] I'm not sure a supportive role of that kind would satisfy me, so I hope it doesn't become widespread for some time. As for writing, the new super-sophisticated AI-powered chatbot released to the public in November 2022, ChatGPT, can easily produce essays, poetry and stories, and surely novels and non-fiction will come[64] – a fellow writer who's well versed in artificial intelligence has seen a sneak preview of the next iteration of ChatGPT and deems it 'unsettlingly good – and it has hints of AGI [Artificial General Intelligence] too, which is alarming'.[65] Open AI's visual tool DALL-E instantly concocts fascinating art and designs according to any parameters you choose, making creatives everywhere quake in their boots.[66] If AGI that can learn and perform any task that a human can is now here, how long will my authorly skills be required?

As for teaching, when I returned to it during the pandemic, I found it different. My students were reluctant to speak or to turn their cameras on. Performing to the camera's black eye was like screaming into the void. University lecturers across the country and world were encouraged or compelled to record or pre-record lectures, generating video and audio content that would become the property of the universities for which they worked.[67] When I failed to record one lecture, I was able to refer a student who missed class to the video I'd uploaded the previous year, which worked just as well.

That gave me pause: by transforming all my mid-career expertise into digitally captured content, could I be putting myself or other people out to pasture, deprived of meaningful employment?

This bit of musing sounds crazy, but maybe it's not. At Concordia University in Canada in early 2021, Aaron Ansuini was enjoying his remotely delivered module on the history of Canadian art. His teacher was excellent, one of the most respected and well-known art historians in the country. Even on pre-recorded lectures, his passion and personality shone through. At some point in the term, Aaron asked Professor Gagnon a question via email, and was surprised when he didn't receive a response. After a bit of Googling, he discovered why. Professor Francois-Marc Gagnon had been dead for two years.[68]

And that made me hesitate even more. At midlife, I've produced and stored so much knowledge, so much work. If someone were to combine all that with a truly artificially intelligent chatbot, something that could not only parrot my content but generate new material, would there still be a place for me? Or would I be left to stagnate on the sidelines?

Is there still a place for me? Midlifers have been pondering that for a long time, but the astonishing pace of technological progress these days may be leading them to ask it even more. Confronted with wondrous inventions that were confined to science fiction during analogue childhoods, it may be tempting to doom-monger, to throw in the towel, to feel helpless and victimised by technology.

But neither Generativity nor Stagnation can be handed to or forced on you. Even in the most difficult moments, there is always a greater range of internal and external responses than you imagine. To be maximally generative, and to evolve with and alongside technology, nurture this mindset: you always have more freedom than might initially appear. On the arrival of ChatGPT, for example, some of my lecturer colleagues declared the death of the essay and moved to other methods of assessment, saying it was now impossible to ask students to write.[69] But others, slightly rattled but not daunted, stayed actively curious about how they and their students could *use* ChatGPT, creating all manner of novel exercises and assignments employing it.

Whether you're a low-PROI Neo-Luddite or a high-PROI Technophile at midlife doesn't matter in and of itself. There's no right or wrong in embracing or resisting technological innovation at this phase of life. What counts is what's workable or not workable *for you*, values-aligned and not values-aligned *for you*, and how honest you're being with yourself about both of those things. When something doesn't feel right about your chosen or imposed relationship with technology, or your connection with others via technology, maybe resistance makes the most sense. When a technological innovation could serve you well but initially threatens or befuddles you, you could instinctively avoid it, or you could take a chance and embrace it. Actual experience with something will always lead you to better choices about your life than speculation, prediction and assumption.

8

Older Adulthood

A cursor waits in the far left of the search bar, pulsing in time with the piano chords against an accompaniment of increasingly lush and heartbreaking violins. The typing starts, tentative: *how to not forget*. The results of the search appear, tips and tricks for improving memory. *Repeat a detail*, one suggests.

'Hey Google,' a man's voice says. 'Show me photos of me and Loretta.' He chuckles. 'Remember? Loretta hated my moustache.' *Okay, I'll remember that*, says Google, with a chirpy beep. The man laughs again.

The music, poignant and swelling, clutches at the chest.

'Remember, Loretta loved going to Alaska. And scallops!'

He asks for photos from their anniversary, requests their favourite movie. All the filmic devices designed to manipulate my emotions are present, accounted for, functioning.

'Here's what you told me to remember,' says Google, as a time- and date-stamped list of his requests scrolls across the screen. Tulips were the flower Loretta loved the most. She had the most beautiful handwriting. She constantly hummed showtunes. She always said to him, 'Don't miss me too much and get out of the dang house.'

And at the end of the advert, he does, after asking Google to help him remember that he is the luckiest man in the world. You hear a screen door swing and slam, and jingling dog tags on a collar.

'C'mon, boy!' the man says, cheerily, as he goes out of the door and into another day of his life.

We are meant to understand that this man is alone but for his canine companion. His warbling voice tells us that he is of advanced age, and his entry into the search box tells us his memory is failing. I imagine him afraid. Without memory to knit together the threads of our experience and learning, the self unravels. If he cannot remember Loretta, the woman who sits at the core of him still, he knows he will cease to be. Although we're never told, we know from the first notes of the score that Loretta is dead.

'A little help with the little things,' says the text in the penultimate frame, before swirling into a rainbow-coloured Google logo that fades out with the music.

Marketing in the United States doesn't get more high-profile than the cinematic, sometimes controversial Super Bowl ads. Many people watching the Super Bowl haven't come for the football at all, but instead to be impressed, shocked or moved by the ads that come onscreen when play on the field stops. Only heavy hitters like Doritos, Budweiser, Coca-Cola, Microsoft and Google show up here; a Super Bowl spot will set a company back $7 million.[1]

Google's Loretta ad[2] is unlikely to have seemed controversial to the vast majority of the 112.3 million viewers of Super Bowl LVI.[3] *Good Housekeeping* gushed over it, posting memes of people soaking their sofas in tears.[4] 'The Google Advert that Tugs on Your Heartstrings', the headline in *Time* read.[5] Lorraine Twohill, the Chief Marketing Officer (CMO) at Google, issued a statement revealing that the story was true and the man real, the grandfather of a Google employee.[6]

Not everyone was so sanguine about the Loretta advert, however, including privacy advocates and commentators including Joelle Renstrom of *Slate*, who cast a cynical side-eye at the company's avowed intentions. 'It's hard to regard Google's supposed interest in helping older people as anything other than dubious,' she said.[7]

And I understand. When it comes to our elders, sometimes alone and often vulnerable, what's the balance between what technology gives, and what it takes away?

Far from seeing the last stage in our physical lives as being about marking time, Erik Erikson viewed older adulthood as a final opportunity to consolidate your identity, and to take the measure of your life. The age of Ego Integrity vs Despair acknowledges that old age is not always kind, and some of us slide into inertia as the sands slip through the hourglass, but there are more positive possibilities too.[8]

The word 'ego' has Freudian, psychological connotations, but in ancient Greek and Latin it simply meant 'I' – the self, the identity.[9] To experience Erikson's Ego Integrity in older adulthood is to feel at peace with yourself and your relationships, content with how your life has panned out in the big picture. Regrets, you've probably had a few, but then again, perhaps too few to mention; you're accepting of and proud about how your life has gone. I often invite my younger psychotherapy clients to inhabit an older person's vantage point to help them clarify their current values. Whether it's 'the ninetieth birthday exercise', 'the obituary exercise' or 'the tombstone exercise', the point is to help people make better contact with what a good life looks like for them by peering at it in the rear-view mirror.[10] If you're able to achieve Ego Integrity when it comes time to close the book, you leave the world with a keen sense of who you are, and with rich memories of a life well lived.

Ego Integrity's shadow side, 'despair', is a bitter pill to swallow, with some older adults having a Monday-to-Sunday pillbox full of it. Little brings with it more despondency than the depressing conviction that your life was wasted, that you didn't do or produce what you would have liked, and that – even worse – you might have reached a point where it's too late or feels too hard to turn things around. For those of us unfortunate enough to develop significant memory problems, the vanishing of our coherent life narrative can cause anguish too.

So, a new developmental challenge has now arisen for seniors, courtesy of today's technology: Coherence vs Fragmentation. For some elders, tech will help them consolidate and maintain identity in their twilight years; for others, it will unearth information that dramatically upsets the Ego Integrity apple cart, deconstructing established life narratives about origins and family and forcing eleventh-hour rewrites of internally held autobiographies. Whether the revelations that come from genetic genealogy will provoke integrity or despair, whether they will cohere or fragment an individual's sense of self, eludes easy prediction.

Technology and psychology combine with older adulthood in one of my favourite comedy sketches.[11] Comedienne Amy Schumer sits on a sofa in a psychotherapist's office. She isn't sure if she can do this, but her therapist encourages her, reminding her that this is about her recovery. Schumer, looking deeply apprehensive, goes to the door to admit her mother.

'Mom,' she says, taking a deep breath, 'what's the issue you've been having with your computer?' Three words flash onscreen with a crash: MOM COMPUTER THERAPY.

'I just think that machines don't work around me,' the mother says. 'They just don't.' I recognise the phrase; my mother-in-law uses it every time she asks me to close the hundred browser windows she's accumulated on her tablet since my last visit. In the sketch, after a triggering series of exchanges, Schumer collapses, foaming at the mouth.

While it may still be hilariously relatable on some level – efficiency and skill with newer technology still differs by generation – the stereotype of the befuddled, Luddite senior has officially reached obsolescence. My mother-in-law might believe technology stops working on contact with her fingers, but she does *have* a tablet, and 86 per cent of online seniors own an average of five devices, on which they spend at least six hours a day.[12] With younger demographics having been tech-saturated for some time, older adults now represent the fastest-growing group of internet users.

Before the pandemic began, the proportion of internet-using over-seventy-fives was burgeoning,[13] but during lockdown even more turned digital.[14] Ultimately, that would prove an adaptive move, enabling them to function better in the more digitally entrenched post-lockdown world. As high-street bank branches close and GP surgeries cope with waiting-list strains through telehealth consultations, anyone who's not online is being shoved to the disadvantaged, isolated margins of society. But 91 per cent of seniors use technology to keep in touch, 87 per cent to organise their money and 73 per cent to look after their wellness and health, a proportion that's set to increase.[15]

The research group providing those numbers, ThinkWithGoogle, has skin in the game, and any internet search about seniors and technology will throw up a panoply of articles about how much seniors can benefit from Google Home, an app that sets up, manages, automates and controls devices within a living environment to create a responsive, connected and potentially safer 'smart home' for elders.[16] This doesn't take away from the numbers, though – whether most seniors are 'enthusiasts', as the ThinkWithGoogle piece says, most of them are connected.

Older adults have always been vulnerable to scams, and now cybersecurity issues such as phishing, ransomware and misinformation are a threat to today's elders, whose more limited technical knowledge and experience can combine with age-related cognitive and physical declines to create a perfect storm. Still, given the potential for tech to improve quality of life in older age, we need to extract the good while minimising the bad. One of the applications of technology that seems firmly on the side of the good is the preservation of memory.

Forgetting has always been the default – on both social and individual levels, we jettison many of our recollections of the past.[17] Even healthy minds lose track, let memories crumble to dust, only selectively holding on to what's meaningful or important. That's natural for all of us. Yet our digital world has ushered in an era where, for the first time in history, it's easier to remember than it is to forget. Smartphones and virtual assistants act as external memory banks, always available for enquiries about what's happened in the past, or about what needs to happen today or tomorrow.

As we age, the need for memory assistance increases. After the age of sixty-five, about 40 per cent of us can expect to experience some form of memory loss.[18] In most cases, this will be relatively minimal; we'll continue to retain strong memories as we age, and we'll be able to hold onto important skills and knowledge. While medical strides will probably be made to reduce this percentage, currently one in five of us will eventually forget ourselves, in the literal sense of the phrase: we will experience dementia.

The terror of the prospect of dementia lies in the fact that memory and identity are linked.[19] We retain a sense of ourselves because we're able to remember our history and the relationships we've had, and we can recall former versions of ourselves, knitting the iterations together into a continuous thread. Dementia snaps the thread, snips it into jumbled pieces, destroys the continuity. Anyone who's been close to someone with dementia knows the pain of witnessing someone unmoored from their history and, therefore, from themselves.

To address retrospective and prospective memory issues ranging from the slight to the severe, memory tech is developing quickly, and the older generation is ready.[20] Over half of over-sixty-fives, including those with so-called 'geriatric cognitive disorders', have smartphones, many of them equipped with adaptations that make it easier for sight-impaired, hearing-impaired users to use them.[21] When I picture the gentleman in the Loretta advert, whom we hear rather than see, I imagine him in a traditionally uphol-stered chair with a smartphone and an Echo Dot nearby, these devices working together to reinforce an ongoing, coherent identity through bolstering his memories of his life and his wife. But today's elders can be far more thoroughly wired up than this.

Imagine an elderly woman ageing in place – staying in her familiar environment for as long as possible. She is surrounded not only by beloved items that spark fond memories, but by tech-nologies that help maintain a sense of self, physically protect them and make independent functioning possible for longer. Picture the house, starting with a digital photo frame on a side table, full of familiar and favourite people, complete with captions and names to reinforce connections that may be loosening. On her smartphone, equipped with various accessibility tools, apps like Eidetic[22] and Elevate[23] help keep cognitive skills sharp. Because music is often so effective in evoking memories, Spotify is also on her phone, its playlists replete with songs Mum knows.

The tech spreads far beyond the phone. Motion-detector illu-minations make up for any failure to remember that when it gets dark, the lights need to go on for safety. Electronic locator tags are tucked into her handbag and coat pockets and dangle from her keys, and a motion sensor near the door can issue a final verbal reminder to take essential things with her. The door itself is equipped with biometric access, an iris scan or fingerprint panel compensating for the possibility of lost keys. If wandering and confusion have become problems, walking out of that door might be a problem too, so an exit could trigger an alert on a carer's or neighbour's device. Smart medication dispensers, stoves and water taps that shut off automatically, and connected, monitored medical and fire-alert systems are essential Internet-of-Things appliances for the older adult.

Such a wired-up home could involve a significant initial and ongoing outlay, but the bill weighs up well against one month in a residential care home in the UK, which could cost thousands a month.[24] Ageing in place can be important for psychological and emotional reasons too.[25] In an Italian study of over 250 elderly people at home, the researchers described a 'choreography' of 'fondness objects' with which the elders made contact throughout their days, objects that provided either memories – like photographs and familiar items – or company, like television or radio.[26] Through these objects they organised their world and supported their strong emotional need for relationship and recollection. And while the jury is out on whether people with dementia do better or worse in assisted living over the longer term, the phenomenon of 'transfer trauma' is well known; especially in the early stages of dementia, people are more at risk of isolation, depression, anxiety and resistance to care when they move to a home.[27] If technology can help people in those earlier stages age in place safely, everyone benefits.

So, the rapidly advancing fields of memory and elder tech are big industries that stand to gift elders and society some big wins, and multiple professionals – including psychologists – are helping design and adapt technologies to make them usable and understandable for older adults.[28] But as is always the case with technology, it can give, and it can take away. The same high-tech kit that helps older adults remember things, stay connected and age in place can be exploited by hackers and scammers. And choices we make about technology earlier in our lives can come back to bite us as elders. For example, research is finding links between sleep and dementia, including an indication that people in their fifties and sixties who get fewer than six hours' sleep at night are a third likelier to develop dementia later.[29] When you consider that 70 per cent of adults use electronic devices in their bedrooms and that the blue light, notifications and distractions that come with it are hurting our sleep quality and quantity, it makes you wonder.[30]

And of course, the consequences of other kinds of choices that we've made earlier in our lives can come back to either haunt or bless us in our older age. That brings us to another major player in the modern technology landscape: the multi-billion-pound genetic genealogy industry.

Jeanne was born in 1959. At sixty-three, she doesn't quite nudge into the 'older adult' range. But this story isn't just about her. Jeanne shows me a book, called *The Girls Who Went Away*.[31] Just sharing the title moves her. The stories in that book, she tells me, echo the story of her biological mother.

'Before free love and hippies and all that happened, before women said, I can have sex if I want, the only thing that really mattered in society was *what will the neighbours think?*' she says. 'Girls all needed to be virgins. So they were sent away.'

Jeanne didn't need to be told that she was adopted. She never fitted in, didn't look like anyone else in the family. She was funny and wanted to talk about everything; they were stiff-upper-lipped, close-mouthed, incurious. They ate meat and potatoes and casseroles, but her body was happy only with vegetables. She fed the meaty casserole to the dog waiting under the table. On vacation, at parties, you name it: she never felt as though she belonged.

At fifteen, she began her search. There wasn't any internet. She and a girlfriend drove to the county seat where she was born, talked their way into the records department, full of big binders. She brought back a phone book from the area. 'Remember phone books?' she asked me. She married a police officer, who printed off a list of every single woman called Virginia Crooks* in the entire United States.

Jeanne shows me the reams of yellowing paper, struck through with highlighter pen in all hues. She made annotations, too, for every phone call she placed. Not once but twice, she rang a number and spoke to her biological mother, who confessed nothing.

'See this top highlight?' Jeanne says, pointing to one line among many. 'That's her.' Written next to the number is Jeanne's note: a date, and the words *not Virginia*.

Around 1989, Jeanne got her first home computer. Immediately she found the adoption sites that there were back then. They were mere connectors, lost-and-found bulletin boards, relying on both parties to put in their information for a match to be made. 'These girls, after they gave birth, they were told to go home and never speak of it again,' Jeanne says. 'And they didn't speak of it again. I knew the chances of Virginia getting on a site, after she'd kept her mouth shut and gone on with her life, were really small.'

She was right. It didn't happen then. But when genetic genealogy came onto the scene, everything changed. Jeanne registered

---

* The surname Crooks is a pseudonym.

with every platform going, including the two biggies: Ancestry and 23andMe, the latter twice, once for the health information they offer and once to find her family. 'When I submitted the second [sample] I came up as an identical twin to myself,' she laughed. Yet one afternoon, not long after that second sign-up, a notification pinged in the corner of her computer.

'You know how you feel when you've lost your kid at the fair, and you think your legs are gonna buckle?' she says. 'It said, *I know Virginia Crooks very well. She's my aunt.* I promptly went on to cry for the next three days.'

Her octogenarian biological mother wasn't keen on being in contact at first – it was a slow process. She was a recluse, an isolated person, a hoarder. As imperfect as it was, she was settled in her life, her story. On her old answering machine, Jeanne has preserved the message from the first call Virginia ever made to her, which was to wish her a happy birthday. 'Sixty years later, she didn't forget,' Jeanne says. Her new cousin, acting as gatekeeper, kept stalling on the initial meeting. 'I'm like, she's eighty-two years old,' Jeanne says. 'Tick tock, you know?'

But then Virginia was dying, and the cousin said to come. Jeanne didn't need to be asked twice. But Virginia was in the ICU, in Covid times, with one daily time slot for one visitor. The cousin guarded that time jealously, but eventually gave way. Jeanne found herself at the hospital's reception desk.

'I straight lied. Well, I didn't really lie,' she says. 'I said, I'm here to see my mom. Which was the truth, kinda. And they said, okay.'

Virginia opened her eyes to see Jeanne, the long-lost daughter that she'd never actually wanted to give away, sitting alongside her. Jeanne was the only baby she'd ever given birth to – she'd never married, never had another child. Sitting up in bed, Virginia opened her arms. *I have loved you forever*, she said. The two women were so similar – their quirky sense of humour, the way they couldn't stop talking, their features.

Jeanne showed me an astonishing photograph of the two of them, holding hands. You can't tell where one hand ends and the other begins, whose fingers are whose. 'There's our crooked front pointer fingers,' Jeanne says. 'See, *this* is what I wanted. That's what I wanted. I didn't want inheritance, didn't want the house. People don't understand. You want a picture of a thumb? Yep. Yes, I do.'

Back in the 1980s, before 23andMe and Ancestry, when there were only phone books and binders in records offices and official printouts from anyone you knew who had access to the right databases, Virginia had been able to deny the truth, not once but twice. But the science couldn't be denied. When Virginia died, Jeanne was there. 'I talked to her until she was gone,' Jeanne says. 'I thanked her for what she did, told her how brave and strong she was. I came in with just the two of us, and she went out with just the two of us. It's the perfect story.'

You'd have to have a heart of stone to not be moved by the tale of Jeanne and Virginia. 'I wish it had happened forty or fifty years ago, but I think she got what she needed before she died,' Jeanne says. To hear Jeanne tell it, their reunion transformed, in the eleventh hour, Virginia's story about herself.

'In the beginning, when I was looking for my biological mom, I thought she would be the missing piece for *me*,' Jeanne says. 'But after talking to her and building whatever relationship we had, I realised that it was my job to close *her* circle. To heal her, so she didn't have to go feeling shame about what she did.'

In the nick of time, the miracle of modern genetic genealogy brought them together. Before her death, with her biological daughter's help, Virginia was able to rewrite her story. They're like the poster mother-and-child for 23andMe.

Virginia's own narrative is one I'll never know. I haven't heard Jeanne's newly discovered cousins tell the tale. I only have the stories Jeanne has woven about her biological mother's internal experience, and about their final moments together: joy, warmth, connection, healing.

There are many potential stories with many potential endings in the world of genetic genealogy. The same sequence of events might be positive or negative, depending on who you're talking to. You order the test, you roll the dice, you take your chances.

Fuelled by a pantheon of commercial services, genealogy has become a well-loved pastime, not quite as popular as gardening but getting there. *Time Magazine* has claimed that genealogy platforms are the most visited type of website after a rather less family-orientated type of site that you can probably guess at.[32] By 2019, over 26 million consumers had taken an at-home DNA test.[33]

In 2020, the private equity firm Blackstone purchased Ancestry.com, the genetic genealogy industry leader, for $4.7 billion.[34] Among the other companies in their stable are drug and medical-device manufacturers, biopharma start-ups and medical research firms. In 2021, they installed new CEO Deb Liu, who'd been with Facebook for over a decade.[35] In 2022, Ancestry announced its expansion into fifty-four new markets across five continents, doubling its market reach. *Forbes* reported breezily that this meant consumers in eighty-nine countries would now be able to 'unlock fun facts' and 'discover relatives they never knew they had'.[36] The Ancestry CEO put a distinctly Facebook-flavoured spin on her vocabulary when she talked about the future of genetic genealogy, saying that Ancestry was all about opportunities, community and 'what binds us together'.[37] By 2024, it's predicted that genetic genealogy will be a $22-billion global industry.[38]

Adoptees such as Jeanne, or the descendants of forced migrants, have reasons for using sites like Ancestry to discover things about their origins that they would never have otherwise known. With little or no access to family medical history, and often struggling to achieve a sense of identity or belonging without their family tree, they're looking to build a historical narrative from scratch. For those taking a DNA test in a more casual way, surprising information that upends their existing autobiography can be difficult to absorb. They might be from a different racial or ethnic group than they believed, might find half-siblings, might discover they're the product of a sperm bank, an affair or a previous marriage. They might stumble into a painful and unanticipated inheritance of secrecy, shame or fear.

Once the secrets are known, they cannot be unknown. The only question is what to do with them.

In Virginia's case, secrecy was enforced and expected, connected to the social shame of a teenager falling pregnant in an era before the sexual revolution and the women's liberation movement. Not all discoveries about parentage are about children born out of wedlock or as the result of affairs, though. In her 2019 book, *Inheritance*, memoirist Dani Shapiro describes taking a DNA test in a nonchalant way, prompted by her husband, who'd become fascinated with his own family origins.[39] Her action revealed a long-kept family secret: her father was not her father, and the donor sperm had come from a medical student at the clinic where her mother was artificially inseminated, something not typically done in their Jewish community. Shapiro got far more than she bargained for. She was thrown into questioning her whole personal and religious identity, and her belonging within her family and community.

Her quest for understanding led her to contact her biological dad – a white-haired retired doctor, a father and grandfather, living on the other side of the country. He got more than he bargained for as well, probably having long forgotten about the donation.

In another example, a mother who had received anonymous donor sperm to conceive her child submitted her daughter's genetic profile to 23andMe. She said that she was wondering about possible health issues and ancestry but was surprised to locate a relative. Her shock was, perhaps, naive – connection with living relatives is an obvious potential outcome of turning over your DNA to a genealogy website. However, Danielle Teuscher didn't think she was doing anything wrong when she contacted the person shown on the genealogy website.[40]

'I think it was the donor's mother that she contacted, and they were maintaining some kind of relationship, because now this mother was a grandmother,' says Jodi Klugman-Rabb[41] from Right to Know, an organisation that advocates updates to privacy laws around adoption and donor conception.[42] When it got back to the donor, he put a stop to it – he didn't want an emotional relationship himself and didn't want anyone else in the family to have one either.

'Why do donors think they have the right to make decisions for the rest of their family?' Klugman-Rabb asked. 'Their parents, now grandparents, oftentimes are interested in some sort of connection or communication. Do the donors have the right to make choices for somebody else?'

It wasn't the only relationship or potential relationship that was severed after Teuscher and the donor's mother made contact. Northwest Cryobank sent a letter, threatening Danielle with a lawsuit for violating the terms of their contract and revoking her permission to use the vials of additional sperm that she had hoped to use one day to give her daughter siblings.[43]

When Danielle was asked by a journalist about the contract she'd agreed with Northwest Cryobank – including the term stipulating she wouldn't attempt to identify or get into contact with the donor – she said she'd never seen it.[44] I'm not surprised. Most of us fail to read or understand the legally binding contracts we sign these days, online and off. In a review of genealogy sites' privacy policies, for example, one journalist found that the shortest was twenty-five pages; another over forty.[45] To obfuscate things further, there were multiple documents to wade through: terms of service, a data-security policy, a privacy policy. You have almost zero chance of understanding what you're signing up for.

Danielle sought the support of Right to Know, an organisation that advocates for updates to privacy laws around adoption and donor conception. 'With over-the-counter DNA testing, there is no such thing as anonymity any more,' says their website.[46]

Jodi Klugman-Rabb became involved with Right to Know because of her own experience. After she took a DNA test in 2014, she hired a genetic detective to track down her biological father. She contacted him and he initially denied her, shut the door in her face. But she went back, determined, and it worked.

Once she was sure they'd have an ongoing relationship, she confronted her mother with the facts. Jodi says that her mother stone-walled, sobbed, insisted that the test was wrong. But it was science vs the story. After a while, Jodi says that her mum claimed she hadn't meant to tell an untruth, that she had never known for sure, that she'd wanted so badly for another man to be Jodi's dad. According to Jodi, her mother had exploited a fragment of ambiguity about Jodi's paternity to support the narrative she herself wanted to have.

'Her refusal to give up that narrative has cut off our relationship,' Jodi told me. 'My relationship with my mother barely exists now.'

The day after I spoke with her, Jodi was travelling to another state to scatter some ashes – the ashes of the biological father she'd only recently met. She doesn't regret finding him. She'd lived with what she calls genealogical bewilderment all her life, the effect of not knowing to whom she was biologically related. When she tracked him down, that went away.

Decisional privacy is the concept that you should be allowed to make up your own mind about who you associate with, who you consider to be your family.[47] Being able to choose that family contributes to the well-being and emotional security of certain types of families, including those where there is surrogacy, adoption or artificial insemination from donor sperm. When all parties reach the same decision, the story has a happy conclusion – individual and familial identities clarify, expand, cohere. When the parties aren't on the same page, when one person's search for connection is another person's intrusion, the end of the tale is different, however.

I think of all those elders on the receiving end of genetically driven challenges to memories, stories, identities. Many have stuck to certain stories for a long time, perhaps so long they came to believe them. Whether they welcome or recoil from the twist at the end of the play depends on so many factors. The older-generation person is usually the audience member in this immersive production, the one who doesn't expect to be called on, to be drawn into the plot.

Genealogy has long been a favourite hobby of the retired, people with time on their hands and an inclination, given their stage in life, to take this moment to chart their place in the broad sweep of history. Even so, they've been slower on the uptake with expensive, newfangled spit-in-a-tube genetic genealogy; note that it wasn't Jeanne's elderly mother or Danielle's daughter's grandmother that submitted a DNA test, but the next generation down.

Although in 2018 only one in ten older adults had done a direct-to-consumer genetic test and only 5 per cent had done so for medical reasons,[48] 60 per cent said that they would *like* to, to estimate their future risk of disease and/or to determine their ancestry.[49]

One problem, though, is that commercial DNA testing often delivers vague, insufficiently specific, ambiguous reports that deal in possibles rather than definites. That can cause more problems than it solves, leading elders down rabbit holes of possibly unnecessary medical work-ups. In a country such as the US, where health insurance is so integral to medical care, this is concerning; perhaps it's more stressful or consequential in somewhere such as the UK, where your anxieties about what's lurking in your genetic profile might be stoked by a commercial test, but the means to pursue the questions that have been sparked aren't available on the National Health Service. Dr Preeti Malani, the co-author of a 2018 study about seniors and genetic testing and a specialist in geriatric medicine, warned, 'Patients may not think about the downstream effects of direct-to-consumer genetic testing. An unexpected positive result may lead to several additional tests that may or may not be covered by insurance.'[50]

And it's becoming clear that insurance *fraud* was becoming an issue for elders too.

In 2019 in California, a disturbing new trend was emerging among the cases reported to the state's Senior Medicare Patrol, which receives complaints about potential fraud. About one in four cases reported that year were related to genetic tests, and on another fraud hotline in the same period, there were now fifty calls a week connected to potential genetic-testing cons, compared with one or two a week the previous year.[51] Something very odd was clearly afoot.

At all sorts of places where seniors tend to gather in larger numbers – health fairs, senior centres and antiques sales, for example

– booths were being set up offering genetic testing to assess risk for disease. They put the scarers on people, talking up their potential vulnerability to frightening maladies like cancer. *You could be at risk for this, you could be at risk for that.* There were other ways they got in touch, too: Facebook, Craigslist, emails, cold calls. Some elders were getting people knocking on their doors, nice folk equipped with test tubes and cotton buds to swab the insides of cheeks, and notepads to take down sensitive information including insurance policy numbers and Social Security numbers, the US equivalent of National Insurance numbers. Sometimes the people manning the stall or knocking on the door or making the call claimed to work for Medicare, a federal health insurance plan in America for people aged sixty-five and older.

They didn't work for Medicare. What they did do, however, was *bill* Medicare for the tests they'd performed – bills that Medicare paid. In return for submitting their biometric information, compromising their privacy and rendering themselves vulnerable through submitting personal data, the unwitting 'patients' got nothing – no reports of results and no further contact.

If the pressures to engage in genetic testing don't originate from scammers and scaremongers at the gates of the retirement community, they might come from within circles closer to home. When you receive a DNA test kit from a family member who's emotionally or intellectually invested in the results, it can be hard to back out of it. 'Give the gift of discovery,' touts Ancestry's website, but this is one unwanted present that's harder to drop at the nearest charity shop.[52]

On NBC's *Think Again*, a group of people explained why they took the test.[53] Most cited health fears, but one man remarked, 'It was bought by my girlfriend, and because it was a Christmas gift, I had to take it.' Dani Shapiro, originally unfussed about such things, might never have discovered her family secrets if the genealogy bug hadn't bitten her husband. Even Al Gidari, the former Consulting Director of Privacy at the Stanford Center for Internet and Society, doles out DNA kits in holiday season. 'It *is* an odd thing for a privacy practitioner to embrace the dissemination of your DNA into the unknown universe,' he admitted to me.[54]

Your communities can also inspire – or insist on – genealogical exploration. The modern passion for tracing heritage dates back to the 1976 book *Roots: The Saga of an American Family*,[55] followed by the blockbuster 1977 made-for-television series,[56] which traced the arc of an African American family from a kidnapping in the Gambia in the eighteenth century, through slavery in the United States, to the present-day author of the novel, Alex Haley. Genealogical research has since become an important way for many descendants of enslaved people to discover their origins 'beyond the wall of slavery';[57] the African Ancestry DNA service advertises itself as being a company by Black people, for Black people. 'Knowing where you're from is a critical component of knowing who you are,' the tagline reads.[58]

Encouragement to discover one's genealogical origins can sometimes be more exclusively than inclusively valenced, though, through focusing on whether someone meets the criteria for membership in a particular 'tribe', whether they are in-group or out-group. As reported by a research study on white nationalism and genetic ancestry testing (GAT), 'Many in the white nationalist and newly emerging alt-right have flocked to the tests and encouraged others to take them' as a way to prove they 'belong' in a group with selective, racially driven admissions criteria.[59] The researchers go on to describe the disruptive and dislocating potential for genetic technologies on both individual and group identities, with the information that's revealed neither good nor bad, but thinking

– or group thinking – making it so. The joy of discovering 'what binds us together', as the Ancestry CEO put it, comes only if the people you're bound to are people you're prepared to accept.[60]

Spiritual communities can have influence too, driving elders to test, or at least allow themselves to be tested. Professor Julia Creet from York University in Canada reported in her documentary, *Data Mining the Deceased*, that the genealogical industry is strongly entwined with the Mormon Church because of a particular point of doctrine.[61] In Mormonism, proxy baptisms can be performed for both the living *and* the dead, enabling the deceased to join their families for eternity, as long as the deceased person accepts the baptism.'[A] departed soul is completely free to accept or reject such a baptism,' states the official website of The Church of Jesus Christ of Latter-Day Saints.[62]

But not just anyone can submit names of deceased persons for proxy baptism. 'All Church members are instructed to submit names for proxy baptism only for their own deceased relatives as an offering of familial love,' the same website explains. Through genetic genealogy, one living person can be a proxy-baptism conduit to untold numbers of as-yet-unbaptised ancestors known and unknown, so the Church suggests encouraging your elders to spit in a tube ASAP. Julia reminds me that the headquarters for Ancestry – once owned by Mormons before its subsequent consumption by bigger and bigger fish – is a stone's throw from the Latter-Day Saints HQ in Utah.[63] The website for Legacy Tree Genealogists, a group of genealogical detectives affiliated with the Mormons, is hyperbolically urgent about the need to test older relatives.[64]

'We need to acquire, preserve and maintain DNA evidence before it is lost forever,' says a blog piece on the Legacy Tree site, emphasising maintenance, preservation and inheritance of DNA as a de facto good, enabling test results to be used long into the future.[65] What they might be used *for* is never articulated, but the importance of future generations being able to 'connect' is presented as incentive enough. *Before it's too late: Testing older relatives NOW*, screams the blog post's headline, the 'now' upper case.

'The disingenuousness of the industry really bothers me,' says Professor Julia Creet. An internationally known scholar in cultural memory studies,[66] she wrote *The Genealogical Sublime*, about the history of family-tree websites.[67] 'There are two moral messages driving the industry: everyone needs more family, and the more family you have, the more connected and healthier you'll feel,' she told me. 'It's a Pandora's box, you have no idea what's going to happen if you track someone down, but the industry still says you can't know who you are without buying the most intimate information about yourself.'

There's an interesting parallel, Julia pointed out, between the assumptions we make in therapy about the unconscious, and assumptions we make about knowing our genetic information. In both situations, we assume that knowledge, and connections we make through that knowledge, will somehow make us whole. Freud argued that we don't have direct access to the unconscious, but that it's the thing that powers how we feel and how we react.[68] It's in us, at our core, but we cannot grasp it, only experience its effects. If we can come to understand what's in it, the promise is that we'll be healed.

The discourse about DNA is the same. It's supposed to be at the root of everything we are, but it too is invisible and mysterious to us. The idea promoted by both the genealogical and the therapeutic industries is that lack of self-knowledge is bad for you and that, by extension, any process that brings you into closer contact with formerly invisible truths about the self will be good.

And this more narcissistically focused benefit – your personal healing – isn't the only good that the industry argues, Julia explains. Ancestry's new owner Blackstone, she pointed out, is involved in pharmaceutical research and development.[69] The notorious Golden State Killer was charged at the age of seventy-two thanks in part to DNA information gleaned from relatives' profiles on GED Match, a user-sourced DNA genealogy site.[70] GED Match was later sold to Verogen, a for-profit California-based forensic genomics company.[71]

'The argument is, by contributing to pharmaceutical research, your DNA is benefiting everyone. By allowing your data to be used by law enforcement, you're contributing to catching serial killers,' Julia says. 'The dominant messaging is about all of those moral goods.'

Julia told me that she has an alert set up for news stories on genetic genealogy. For every twenty stories, she says, roughly nineteen of them are about solving cold cases, the type of story for which the public has a seemingly insatiable appetite. Only one story in twenty is about privacy. Only once in a great while, she says, does someone pipe up and say, *Hold on. If all these people are using your data, shouldn't you be paid for it?*

Instead, the payment streams flow the opposite way. On commercial genealogy websites, four-fifths of users opt into their data being aggregated for use in medical research, voluntarily forking out over £150 or so for a test and also granting permission to sell aggregated genetic data now or in the future, increasing corporate profits.[72] It's an incredible business model, selling people's biological information back to them, as well as innumerable other parties.

The power and knowledge differentials are stark, and the economics concerning, especially when you consider the potential unanticipated psychological, social and financial costs people risk when they submit their biometric data to powerful information engines stuffed full of family trees branching from here into seeming infinity.

Again: you roll the dice; you take your chances. The test kit arrives. You open the box. And then, Pandora, you open *the* box.

DNA is so intimate that it's hard to believe it's safe to just put it out there, difficult to fathom how such an action wouldn't constitute an individual threat. Privacy advocates bang the warning gong, sometimes in creative ways. For example, an activist group called EARNE\$T advertised an auction of items collected from the 2018 World Economic Forum meeting in Davos.[73] To raise awareness of surveillance capitalism, and to underscore to the public that Google, Facebook and the like want to collect as much of our personal data as possible, members of the group squirrelled away DNA-carrying items including hair, cigarette butts and drinking glasses from the world's leaders. In the auction catalogue, EARNE\$T said that successful bidders would be able to 'gain key insights into Davos attendees, including . . . potential physical and mental ailments or gifts, ancestry, records of diet and medications, substance abuse and exposure to environmental factors'.[74]

Among the items being put up for auction was a drinking glass allegedly bearing the DNA of Donald J. Trump.

There's a feeling in the pit of your stomach that corporate or state access to such intimate information, the DNA-related data of vulnerable parties such as children or elders, just has to be bad, bad in a way that sits in the pit of your stomach.

But Al Gidari, the top-tier privacy expert who gifts DNA kits for Christmas, feels he has good reason for nonchalance. The genie, he says, has been out of the bottle for a long time.[75] So much information is out there about us, he says, and the algorithms are so powerful, that you don't *need* DNA to gain any insight you want [about a person], to act prejudicially against anyone you want. The accumulated health and other data that's already available is enough: physiological information from your wearables, data you've freely offered up yourself, health information from people that the internet knows are biologically related to you. While you can get het up about your genome being out there, the thing that could end up denying your health or life insurance could be other data within your digital footprint that's been collected via other means.

'Sometimes DNA gives you a quicker answer, confirms a fear, or alleviates one,' Gidari says, 'but you [already] know from the external facts that there's a [health] risk.' Your DNA may reveal an increased danger of a heart attack, but that's hardly a shocker if your existing data footprint has already let that cat out of the bag. Your existing data profile may reveal a panoply of risk factors: your obesity, your lack of physical activity, the two packs of cigarettes you buy every day, cardiac data from your smartwatch, your regular orders from local takeaway joints and the incidence of heart problems among your relatives.

So, as a family-history enthusiast himself, Gidari carries on wrapping up those DNA-test gifts for Christmas.

He sketched me his overview of the life cycle of so many of our fears about new technologies. 'First, we gasp in horror,' he tells me. 'Then, there's the parade of horribles. Then, we have the moment where it becomes normalised. Then we realise the bad things didn't come to pass. And then there's version two, resurface the prior discussion, and go through the whole thing again. The reality is almost none of the bad things that people fear have come to pass on any technological innovation.'[76]

For all our sakes, I hope he's right.

Erikson said that the existential question for the Ego Integrity vs Despair phase of our older years was 'Did I live a meaningful life?' Note the form of the verb: *did*. For modern older adults, living in such a technologically and informationally infused world, the verb tense seems all wrong, too static. The pace and reach of techno-logical progress mean there's far too much still going on to be thinking in the past tense – for better or worse.

Technology brings identity coherence when it helps a senior preserve memory; consolidate and maintain their sense of self; and stave off despondency and loneliness. As portrayed in the 'Loretta' advert, virtual assistants can help elders with reminders and reinforcements to strengthen their positive memories and live independent, functional lives for longer. But tech's potential ability to preserve and enhance both retrospective and prospective memory is only part of it. If an elder wants to go on a quest to consolidate a feeling of identity and community that has always eluded them, there's never been a better time. Because of the expanding net of genealogical connections encircling the globe, those searches will only become quicker and more straightforward.

But when the results come in, or when the call or knock or message you didn't expect arrives, comfortable assumptions about history and identity can blow to pieces. Rewriting your stories, redefining yourself, and rejigging your relationships aren't easy pro-cesses when you're set in your ways and have lived with one set of supposed truths all this time. At least the research shows that, after the relative fixedness of middle age, our personality traits loosen again in older age, becoming almost as malleable as they were when we were children. Perhaps that can make it easier to accept change, to stay open to experience and possibility. But for older adults, or indeed for anyone at any age who decides to turn genetic detective, this technology could take you in either direction: coherence and consolidation, or fragmentation and revision.

Digital technologies are transforming and may yet transform older adulthood in so many ways that have not been explored at any length here. When mobility and or technological skills are limited, today's communication gadgets could keep seniors far more connected with peers, services, communities and other generations – vital when, at the moment, 1.4 million older people in the UK describe themselves as 'often lonely'.[77] How soon might the same gadgets that now act as portals to friends and family actually start serving the same *functions* as friends and family – robot helpers and companions providing care and friendship at times it's not available from elsewhere? How soon might the dividing lines that exist between human companion and artificially intelligent companion start to fade?

These musings still seem futuristic, sci-fi-like, but developments such as this might not be as far distant as you think. And this takes us up and over the edge, into the *new* final stage of the modern human life cycle.

9

Digital Afterlife

Each year, Amazon holds a global event showcasing technologies on the cutting edge of the future. Attendees come to re:MARS to be inspired and titillated by scientific and thought-leader luminaries touting the latest innovations in Machine Learning, Automation, Robotics and Space.[1] Chief technology officers, chief executive officers and chief scientists pace the stage against the backdrop of a massive screen displaying high-definition demonstration videos.

In June 2022, Amazon's Senior Vice President and Chief Scientist Rohit Prasad addressed the audience, demonstrating how AI could make our lives ever better in the years to come.[2]

A woman and a small robot with a tablet for a face stand companionably together, studying the sliding-glass doors in a lovely mid-century modern house.

'Astro,' the woman says, 'this is the backyard door.'

Astro gives this portal to his new home the elevator-eye treatment with his tablet face. Later, we see him wheeling idly about the house. Hesitating at the door from before, he snaps a photo and wheels to where the woman is reading, blissfully unaware her perimeter may be insecure. Displaying his photo, he presents a text question on his face.

'Yes, the door is open,' replies his . . . what? Is she his employer, his owner, his friend? 'I'll go close it. Thanks, buddy!'

Buddy. Perhaps friend, then. In this keynote speech, Prasad's keen to stress the potential for us to have human-like, emotional, 'real' relationships with digital things. Empathy, affect and trust, Prasad argues, have become yet more essential during the pandemic, when so much has been taken from us. Perhaps innovations like what we're about to see, he says, will make those losses easier to bear. The video begins again.

A smart speaker, shaped like a panda's head, sits in a child's bedroom. 'Alexa,' says a little boy, 'can Grandma finish reading me *The Wizard of Oz*?'

'Okay,' says Alexa, in her familiar voice. But then, the cadence transmogrifies into that of an elderly woman, reading about the Cowardly Lion's newfound courage. A happy if slightly wistful smile sweeps over the boy's face as he follows along on his paperback copy. Perhaps it's a book his grandmother gave him; maybe they used to read together, cuddling before he went to sleep.

Prasad is never explicit about where the grandmother is now. The reference to the pandemic's losses has made the context plain enough. This grandmother, sharing *The Wizard of Oz* with her little grandson, is a voice from beyond the grave.

'We are unquestionably living in the golden era of AI,' Prasad concludes, 'where our dreams and science fiction are becoming reality.'

As he leaves the stage to a patter of applause, watching journalists are consulting their thesaurus app to find synonyms for 'creepy' and are scrolling through their source databases to find experts to interview on this subject – like me.[3] Perhaps those experts will speak not to whether this is possible – Prasad has made clear that it is – but whether such a thing could ever be considered *desirable*.

Much of Prasad's talk didn't make headlines. A cute personal robot telling you the back door's open might sound reasonable enough to most people these days. But as far as many observers were concerned, the angle on a story about dead people reading bedtime stories to children had to be that it was a Super Bad Idea.[4] When Prasad claims in his presentation that Amazon's customers wanted Alexa to have 'a skill for every moment', *surely* this wasn't one of the moments those customers meant?

Having begun with digital gestation, we end our exploration of the modern life cycle with digital afterlife: the other end of the modern identity. For two decades I've followed what happens to our data when we die and tracked movements in the death-tech space, and it's clear that the place of the dead in our social world is changing radically and rapidly. The amount, richness, persistence and accessibility of deceased people's data is exploding, and new technologies are constantly emerging that could allow us to do more things with that data.

It's becoming easier to imagine that, one day, the digital dead might form a whole new social caste: empathic, humanoid, artificially intelligent entities with whom we can continue to interact.

If the concept of digital afterlife freaks you out, you're not alone. If you think the idea is nonsense, a non-event, you're not alone either. People who map psychological and social life cycles have, perhaps understandably, been less interested in digital afterlives. The clue is in the name: *life* cycle. The non-living haven't traditionally enjoyed legal personhood, consciousness or an active social life. But I'm someone who believes that what happens to our data after death is critical for all of us to be thinking about, and in this chapter I'll describe why I think our digital afterlives constitute a continuing social identity, deserving of its place in a new model of the modern human's lifespan.

Accidentally or on purpose, will your online 'self' persist or disappear after death? If you remain, will others experience you as an empathic digital being with emotions, as Prasad predicts will be possible, or simply as an archive, a dataset? And who will get to decide the form you take? You, as part of your estate planning for your digital material, your loved ones – or a corporation? Through your own efforts or that of others, will you end up as a strong and present Posthumous Influencer, or a wispy Digital Wraith?

Welcome to the new last stage of life, unthought of by Erik Erikson, where you don't shuffle off this mortal coil until you're deleted from the Global StorageSphere.[5]

Try asking someone whether their digital remains will persist or disappear after they physically die. They'll probably look at you as if you're a little peculiar. First, you've dared to bring up not only death but someone's *own* demise. Second, the phrase 'digital remains' sounds weird, not something that crops up in everyday conversation – but that may change.

Digital remains go by a lot of names – digital legacies, digital assets, digital dust, digital zombies – and their existence isn't a surprise, when you think about it. They're the logical extension of our online lives in a system that's not well set up to deal with the fact that users stop being alive. When people ask me what happens to your data when you die, they're uncertain what word is correct for posthumously persistent digital information attached to a particular person. Digital *stuff*, as in something that's owned? Digital *self*, as in something you are? Digital *footprint*, as in something you leave? Digital *afterlife*, as in something that continues to be vital online, unleashed from the carbon-based creature that spawned it?

Whatever the best word for digital remains, one thing's for sure: they are legion. Obviously, there's tremendous variance in the size and comprehensiveness of digital footprints among individuals, the gap being widest between the citizens inhabiting the most and least digitally connected parts of the world. Simply put, the more data you generate in life, the more you'll leave behind in death. In a blessing for our planet's struggling environment, not all those energy-sucking data are kept, but a considerable amount will stick around by design or default, at least for a while.

Your stuff is in the Global StorageSphere right now, of course. Consider how much of it you stashed in the cloud to declutter your office of paper or free up storage on your devices, those thousands of files you've created and never get around to purging. After all, passively preserving your information and letting it sit with Apple or Google or Dropbox is far less trouble than sifting through and editing it down. You might need it sometime. What would your motivation be to delete it if it's not hurting anyone?

Yet, as I write this, the UK and much of the world faces a climate and energy crisis.[6] From this vantage point, it's sobering to realise that our digital lives will soon account for a fifth of the electricity used globally and that data-storage centres will use more than one-third of that.[7] By 2025, 'the cloud' will probably contain over 200 zettabytes. The amount is hard to fathom, but if you own a 1 terabyte hard drive, a zettabyte is equal to 1 billion of those. These data, of course, aren't literally in the clouds but very much on earth, cooled down 24/7 by equipment that will, it is to be hoped, become more energy efficient and carbon neutral as technology evolves.[8]

Nestled within all those servers are the data of billions of dead people. Some of those digital remains are consequential, some mundane and meaningless. Some were generated by the famous, notorious, or whomever we deem historically and culturally important, and some were created by ordinary people who never made a headline or left their home towns and lived both brilliant and banal lives. But whatever the significance or scope, every wired-up citizen of the world leaves a dataset behind: a far-flung, multimedia, occasionally coherent but mostly fractured feed of information that trails behind you like billions of drifting, cast-off skin cells.

Some online dead are clearly gone from the physical earth, while the status of others is unclear, or something we don't consider when we're affected by a stranger's words, images or videos online. You may have been moved to book your last beach vacation based on a dead person's TripAdvisor review. And let's not forgot those who lived and died before the digital era, but whose images and narratives are endlessly storied and restoried, sold and consumed online. Today's kids don't know Marilyn Monroe from the movies or Kurt Cobain from concerts: they recognise these cultural icons from the internet.

In the physical world, dead bodies either disappear completely, their ashes ritually scattered to the four winds, or they're swiftly and clearly segregated into their designated areas: cemeteries, grave-yards, columbaria, urns. Although up to six in ten of us experience occasional hallucinations after losing someone we love – hearing their voices, catching a glimpse of them or feeling their touch – we don't run into them at parties, encounter them in our houses or have two-way conversations.[9] If we do, we generally call it a haunt-ing, hardly an everyday experience.

With the digital dead, it's a different story: we *do* encounter them daily, whether we know it or not. Although 'virtual cemeteries' exist, dedicated places for online memorials, the data of the dead generally occupy the same places and spaces as the data of the liv-ing, because there's no one to organise it otherwise.

Nowhere is that more visible than on social media. Inspirational vloggers and influencers remain on YouTube or Instagram after they've succumbed to cancer or addiction, suicide or homicide, continuing to rack up followers. Researchers at the Oxford Internet Institute (OII) have calculated that Facebook will be transformed into a cemetery of over 4.9 billion digital graves by 2100, the scales tipping in favour of the dead sometime in the third quarter of the century.[10]

The dead of Facebook are proliferating because posthumous profile deletion is opt in, not opt out. If you want your profile to be deleted on death, you must take specific action, delving into your legacy-contact settings to tick a box. If you don't, your legal representatives on earth might or might not bother to take action to remove it. If you don't think ahead, you'll stay socially networked after you're gone.

When it comes to planning for our own deaths, doing nothing is the default setting. Only about 30–40 per cent of people in the US and UK have a will for their *physical* stuff; the percentages of people who've attempted to make provision for their digital estates is considerably lower.[11] When people *think* about what they want, they tend not to action it. In the 2018 Digital Death Survey, about 31 per cent of respondents said they were likely to request their social media be 'removed' after their deaths, but only 11 per cent had taken the handful of seconds necessary to set a Legacy Contact for their Facebook account.[12]

Somewhere along the line, many of us have started looking to social media companies to care for us in our grief, expecting them to carry on housing the digital dead. When they try to do otherwise, we protest. In late 2019, Twitter was seeking to release expensive storage space and announced an imminent cull of inactive accounts, which would have included the non-posting dead.[13] Immediately, bereaved users fought back against the threatened deletion of their loved ones' information, causing Twitter to apologise and backtrack within twenty-four hours, promising to investigate memorialisation as some of its Silicon Valley neighbours have done.[14]

I wonder how long social media companies will be inclined to keep this up. While there might be a variety of ways companies can monetise the data of the dead, hosting them in perpetuity is a drain on the coffers. The deceased are sluggish about buying stuff, and they almost never click on ads.

Maybe you don't do social media, but don't think that means you won't have a digital afterlife. Modern life generates plenty of digital remains that are less visible, and perhaps more telling for it: digital documents and photos to be lost or found, scraps of memorabilia to trouble or comfort, threads to weave new narratives or challenge existing ones. Blogging platforms, email and cloud-storage accounts that cost something will deactivate eventually when the bills aren't paid, perhaps taking practical and sentimental material with them, but free accounts stick around until someone figures out how to access or cancel them.

Some of the electronic traces of the dead are more straightforward to access: they sit with us, in our own devices and accounts. Our relationship with them isn't always easy. Message threads with now-dead friends and relatives stay on our phones, slipping down the list of conversations unless we message them, which many of us do. Deleting a deceased person's contact from a phone's address book may feel complicated, part of the fraught territory of modern technologically mediated grief.

We tend to underestimate the power and influence of the scattering of bits and pieces we leave of ourselves online. Whether you're alive or dead, anyone can come along and stitch together a *bricolage* from those pieces, narrating you and your life however they choose – often getting quite close to what you might have once thought of as the 'real' you. I've demonstrated the power of this capability many times, in workshops where I conduct an exercise called 'Eulogy for a Virtual Stranger'.

At these events, someone with a manageable level of death anxiety and a substantial and accessible digital footprint offers to be eulogised. A small team of eulogisers are given only an initial entry point – a LinkedIn profile, a website or an Instagram account. Their task is to construct a heartfelt, personal tribute in only half an hour, capturing the essence of the person's existence, who they 'really' were, and what was most meaningful to them in life. Diverging from the usual practice at funerals, after the presentation of the eulogy the 'deceased' has right of reply, feeding back about how well the eulogisers captured his or her essence.

The results are always the same: the eulogised volunteer is floored. Not only are the details usually largely accurate, but the overall sense conveyed reflects how they see themselves or want others to see them. Save for the odd incorrect detail or error in emphasis, the astonished volunteer says *yep, that's me.* The other element of surprise is how many details in the eulogy they don't recall putting out into the world. Of course, we can't possibly remember or track everything we've revealed about ourselves online over the years, but the eulogisers often drew heavily on things third parties had recorded, stored and shared online, or passively recorded on platforms such as Spotify.

Huge chunks of our eventual digital remains will be information about us that other people have disseminated about us, without our control and sometimes without our knowledge, and the embroidery of others will continue after death, too. While you're alive, you have the chance to correct the record. After you're dead, all bets are off.

As technology preserves our data and makes it accessible past the point of our physical deaths, it's clear that people will continue to experience and relate to us through that data. For many in their loss and grief, that will be a valued, marvellous thing. For the family historians and genealogy enthusiasts of the future, these data may be a wonderful boon. That thought might comfort you. You may feel less sanguine about the fact that both friends and strangers could access, edit, narrate, use or abuse *your* digital remains for their unique purposes and motivations, perhaps reconstructing your identity and legacy in the process.

Later in this chapter, we'll look at ways in which your social identity might continue online in quite vivid ways, either by your design or other people's. But sometimes those left behind won't want to use your data for any sort of elaborate continuation or reconstruction purposes – they might wish only to access material, to settle unanswered questions, assuage feelings of grief or experience a connection with you. But the material we leave behind is often so comprehensive, so private, and so connected with other living persons that the question of whether these data should be protected or passed on to heirs is becoming increasingly sticky.

In 2018, a judge in the highest court in Germany decided that our digital remains should simply go to heirs, like papers in a hanging file folder or a box of photographs. In what proved a controversial ruling, he reviewed the case of grieving parents who were unsure if the death of their daughter had come about through accident or suicide. They hoped the contents of her Facebook Messenger account would give them the answers, but Facebook had refused access. Pointing out that it had always been common to hand over private correspondence, journals and the like to the legal heirs after someone's death, His Honour saw no reason whatsoever to treat digital data any differently.[15]

Amateur genealogists agree. I once spoke about digital remains to a gathering of people who were fervent in the pursuit of their hobby. With consternation, they argued my point that perhaps some data should be kept private after death, especially when intertwined with the information of the still living. If they left it behind, it means that they were all right with it being seen, protested one family-history enthusiast, whose own sense of identity was so intertwined with knowing as much as possible about her forebears that the idea of not being able to access this material was unthinkable.

In family members' reactions to not being able to access loved one's data, you see again and again the theme of our identity being socially constituted, inextricably bound up with and defined by our relationships with others. Their motivations for wanting to see the archive are as much about their own identity as the deceased person's. The German parents were astonished when Facebook denied them access – they were *her parents*, she was *their daughter*. Carol Anne Noble from Toronto found herself unable to carry out her husband's dying wish, that she finish the book he'd begun, because it was stored in an iCloud account to which she'd lost the password; the idea that she could not fulfil her beloved's request was agony – she was his *wife*, she said, incredulous.[16] At first, Facebook refused to allow Gloucester-based Nick Gazzard to remove the seventy-two photos of his daughter's killer from her profile. He had the password, but the account had been memorialised – no one could log in, and nothing about the profile could be changed. If Hollie had appointed a Legacy Contact on Facebook, that person would have been able to make changes, but this innovation was introduced only within a week of Hollie's death.[17] The family could not abide a lasting online legacy showing a much loved daughter and sister in the arms of her murderer.

And sometimes digital remains are connected not just to the dead person but to those who mourn them: family photos, intellectual property that should pass to the heirs or creative outputs that belong to other people. One woman I interviewed for my last book lost much of her photography portfolio when her boyfriend's family took the laptop on which both had stored their respective bodies of work.[18] Before Apple introduced 'Legacy Contact' settings for Apple IDs, Rachel Thompson from London waged a three-year legal battle for access to family photographs in her late husband's iCloud – he had captured their daughter's entire childhood with his photographs, and without them, the little girl would lose this record of her younger years.[19]

But sometimes the desire to plunge into the archive is indisputably about the attempt to connect with a still-beating digital heart, to continue the bond between the dead and the living. We often experience a searching and calling reflex when we've lost someone, looking for our missing piece. We quest after a visceral sense of them. We want understanding, we follow clues, we look for answers. We may want more, more, as much as we can get, perhaps not sure exactly what we're looking for, or when enough is enough.

As static it may be in some senses, viewed and interpreted through the eyes of someone searching like this, the archive doesn't feel dead at all – especially when you consider what's in it. If you're after intimacy through an encounter with the intimate, there is no shortage in the archive, and what we find there might transform our visions and narratives of who that person was and our relationship with them.

Let's say that people can see your archives after you are struck down by the proverbial bus. If that's because of your poor data-security practices in life, their access could be unfettered. If you're among the low percentage of people who've planned for their digital remains, it will be somewhat less comprehensive but still revealing.[20]

If my number comes up tomorrow, my iPhone, Apple Watch and MacBook Pro will be among my personal effects, if they haven't been destroyed in the accident. If they have, that's okay: my ageing iPad and an iMac are at home, and everything is connected to my iCloud account. I also have an external 2TB hard drive, approximately 15 USB sticks, and a motley crew of legacy hardware in the crawl space, most of which you'd need to take to a Museum of Dead Technology to extract any data.

Because I am a psychotherapist, a profession where confidentiality is key, anything I use for work is armed to the hilt with encryption and password protection. I haven't enabled Google's Inactive Account Manager, but my assistant has the information and access necessary to pass information to my 'clinical executor', the fellow professional who would manage my clients in the event of my death.[21] My family wouldn't be able to log in to most of my online accounts – for security reasons and in consideration of other people's privacy, I'm not going to leave a password list lying around – but I've made a list of what they are so they can be closed. My husband is my Legacy Contact on Facebook, where I've given him permission for him to download an archive of data for anyone who wants it; in any case, the stack of books I created before I purged my history are in my study.[22]

Crucially, because of how I've prepared, my Apple Legacy Contacts have access to photos, notes, mail, contacts, calendars, reminders, iCloud messages, call history, iCloud files, health data, voice memos, Safari bookmarks and my reading list. Think about that, really think about that.

Now think about what else can be accessed if someone has all your passwords, access to all your devices, access to everything. That includes the list of websites you visited at three in the morning, and the search history of things representing whatever you were thinking, worrying or fantasising about. The latter is such a treasure trove of insights into the inner workings of your mind that a series of mini-movies, 'I Love Alaska', were made based solely on one anonymous Texan woman's search history, the drama of which unspools like a time- and date-stamped Russian novel.[23] Watch it, and wonder what it might be like if her grieving husband, rather than a couple of film-makers, had accessed this search history on her laptop after her death. What might he have understood about his wife and their relationship that he hadn't grasped before?

And what about location data? If they're enabled or not disabled, the location settings on our phones and wearable devices have the capacity to track everywhere you've ever been, which isn't always the same as where you said you were. Those data are easily triangulated with other passively captured, time-stamped information. When your heart-rate data combines with your unexplained presence at a Welcome Break off the M4 on a particular Wednesday when you emailed your spouse complaining of a boring all-day meeting in the office, a certain picture begins to emerge.

Grievers aren't always tempted to come over all Hercule Poirot on their loved ones' digital legacies, but in situations where the relationship was troubled, or where the death was complicated or mysterious, a bereaved person's hunt for answers to persistent questions can send them into a spiral of compulsive digital detective work. Has there ever been a time when you weren't where you said you were? Maybe it was entirely innocent. Maybe it wasn't. But what might it be like for someone you love to be wrestling with that question after you're gone? How might it affect their conception of you, and of themselves?

When it's comfort you're looking for, you run the risk that your memory of a dead loved one, and your ongoing feeling of a bond with them, might be turned upside-down by a too close encounter with material you didn't expect. Searchers can never know ahead of time what's under the lid.

Privacy lawyer Al Gidari of the Stanford Center for Internet and Society remembers a case like this, one of the first he litigated in this area.[24] A young man of eighteen was riding his motorcycle to college, had an accident, and died. At the time, Gidari says, there weren't social media per se, but there was still a lot of material his parents wanted to access – emails and other digitally stored content. Gidari told them the law – that the contracts their son had with these companies didn't allow them access – but they didn't understand and wanted to go to court.

'It turned out there was a way to help them,' Gidari says, '[contacting] everyone who received communications from [him] and being able to go to them to get permission or to subpoena it. But it turned out very badly because the son wasn't who they thought he was. A lot of the information that they found out, that became their last memories of him, was very disturbing.' To make matters worse, new information about living people surfaced too – perusing both sides of her dead son's correspondence, his mother discovered negative things her daughter had said about her, causing grievous conflict within the remaining family.

'It became a metaphor for me in this whole practice, which was be careful what you ask for,' says Gidari. 'You might not like it if you get it.'

I recall Rachel, a grieving mother I also interviewed for my last book, who accessed her daughter's laptop after her suicide.[25] Despite her best efforts, she hadn't been able to persuade Facebook to change the profile picture, which showed Katie in the room where she died. This loss of control had felt traumatic, and her upset increased when she found additional disturbing images on Katie's computer.

'It was very painful,' Rachel said. 'I didn't know what I'd find on there. Tens of thousands of photos of herself, selfies of herself, looking at herself in the mirror. Looking sad, the next one crying, looking at herself. It was heartrending.'

The archive might not be alive, might not render the person still alive, but, for better or worse, our interactions with it can make it feel that way. And where your discoveries shift the narrative of that person's life, and your relationship with them, it's amazing how vital that can seem.

I've talked about the archive at length because, of all the types of digital afterlife, it's the one you've most likely encountered – perhaps multiple times now. But the technologically enabled life-after-death scenarios you've seen portrayed in sci-fi series such as *Black Mirror* aren't the stuff of fiction any longer, and the possibilities for digital afterlife that exist right now might surprise you.[26]

The dead might not be able to remain sentient through digital technology (yet), and (as far as we know) they're not aware of continuing to be socially interactive and influential, but living humans and their algorithmically driven machines can partner to create an illusion of interaction. If these phenomena start happening at scale, it will have major implications for how we live, and how we conceptualise death, perhaps coming to see physical, organic death as something separate from the death of the self.

At the bottom of the digital-zombie scale are voice-from-beyond-the-grave phenomena, both unplanned and planned.[27] Courtesy of profile-cloning scammers, the dead send out new friend requests. People still logged onto their deceased loved one's social media profiles, often with good intentions, make posts or send direct messages from within those accounts, their communications materialising as though from the dead person themselves.

Benevolent intentions aren't universal, of course. *Don't you think it's a bit soon to fall in love again?* wrote one bereaved mother to her late daughter's newly loved-up boyfriend – from within her daughter's Facebook account.[28]

The more time that passes, the higher the likelihood that an email account will be hacked, so if you haven't yet been spammed by the dead you may be in the minority. Mobile numbers are reallocated, as one young woman realised when, after six months of finding comfort in texting her late brother, she got an answer.[29]

And finally, in the planned department, numerous digital-legacy companies tap into our death anxieties, offering to help us be there without being there through services like pre-recorded, post-humously delivered video or text messages to be drip-fed out over the years on special occasions. Whether the messages will be welcomed by the recipients when they arrive is impossible to predict.

In these cases, the illusion of life, of ghostly hands reaching out to touch us, is fleeting; it generally becomes rapidly clear that there are disappointingly prosaic explanations for the phenomenon rooted in this realm, not in the great beyond. But there are other ways of using the raw digital material available to represent and revive the dead in ways that range from clunky to undetectable.

On Kim Kardashian's fortieth birthday, Kanye West arranged a 'special surprise from heaven': a hologram projection of her late father, Robert Kardashian.[30] Clad in a light-coloured suit, speaking in his own voice, Robert reminisces about how they used to listen to music in the car, doing a little dance to a favourite tune. He says mischievously that he sticks close by and watches over her, his presence signalled by someone passing wind. He lionises her husband, gushing that she has married 'the most, most, most, most, most genius man in the whole world, Kanye West'. Anyone in the room still uncertain about the authorship of the hologram's script had their doubts reduced at that point. (In unrelated news, Kardashian filed for divorce four months later.)

The birthday gift's cost wasn't disclosed, but being of a comparable quality to the Tupac Shakur hologram that played Coachella after his death, it could have been between $100,000 and $400,000.[31] Like the Roy Orbison, Whitney Houston and Maria Callas holograms that have done the circuit in the last few years, the illusion was slick but clearly identifiable as not being the real deal.[32]

The same is harder to say about ABBA's 'Abbatar' performance, in London as I write this. Produced at a cost of $175 million, it is one of the most expensive musical extravaganzas ever staged.[33] Unlike the previous examples, it was done with the full knowledge, participation and consent of the band members, all of whom are still very much alive; there is no doubt in my mind that *ABBA Voyage* will continue to tour and make money after that changes. Unlike the previous examples, if someone didn't tell you what you were looking at, you wouldn't know, both due to the quality of the production and the fact that its subjects were alive to participate in it. You'd think you'd taken a time capsule back to the 1970s, that you were seeing ABBA in the flesh. On the way out, my daughter asked if the band occasionally present at the left of the stage was actually there. Unable to trust my senses, I told her truthfully that I wasn't sure.

In a final instance drawn from the world of entertainment, a whiff of unsavouriness, scandal and dicey ethics hung about the

2021 release of *Roadrunner: A Film About Anthony Bourdain*.[34] The author, chef and travel documentarian died by suicide in 2018 but appeared to provide part of the voiceover for the film.

In an interview with *The New Yorker*, director Morgan Neville disclosed and defended his decision to use AI to create this audio, which Bourdain never spoke, and he went further.[35] 'You probably don't know what the other lines are that were spoken by the AI,' Neville said, 'and you're not going to know.'

The executor of the estate, Bourdain's ex-wife, was unimpressed and denied she'd been consulted. 'I certainly was NOT the one who said Tony would have been cool with that,' Ottavia Busia tweeted.[36]

As a documentarian himself, with a popular show, there was ample voice data from Bourdain to train an AI.[37] Neville hired a professional software company to do that for him, as ABBA enlisted George Lucas's Industrial Light and Magic for *Voyage*[38] and Dr Dre hired Digital Domain to create Tupac's holographic reincarnation.[39] All this involves big money. On the other hand, Rohit Prasad boasted onstage at re:MARS that less than a minute of grandma's audio was sufficient to allow her to deliver posthumously the whole of *The Wizard of Oz*, a technology that would be available to everyone via Alexa.[40] And if you have $30 and 30 minutes available, you can clone your own voice for perhaps an even more convincing illusion.

I already used Descript[41] to produce impressively accurate transcriptions of interviews for research and writing, but it was on *The Future of Everything*, a Netflix series, that another application of the platform became clear.[42] Descript's AI-powered 'Overdub' feature enables 'ultra-realistic voice cloning, [which] lets you create a text-to-speech model of your voice'. Once you've read out a 30-minute script, you can create audio simply by typing out your text.

On *The Future of Life After Death* episode, thanatologist Cole Imperi – a scholar of death, dying, grief and loss – types, 'Thank you for attending my funeral today. I lived a good life.'[43] She wells up. 'I don't know why that's making me emotional,' she said. 'I sound like myself. That is weird. I feel weird.'

Recalling that I trained my own Descript voice clone two years ago, I open a new project and type a playful deathbed confession. 'It was I who killed Colonel Mustard with a lead pipe in the scullery,' I write. Several seconds later, the audio waveform appears, and I hit play. I suspect that nobody, listening to that clip, would know that I hadn't spoken those words.

I notice that Descript offers another feature: the ability to share your voice clone with someone else. For reasons that should be crystal clear in a time of ever deeper deep-fakery and post-factual facts, I'd advise you to think carefully before doing that. As I write this, algorithms trained to detect deepfakes are only about 65 per cent effective; only 20 per cent of deepfakes can be manually identified by humans.[44] And dead people, being dead, have no right or ability to set the record straight.[45]

If your mind isn't blown yet, let's talk about chatbots. Chatbots of the dead have been around for a while, the province of people who know the technology – programmers and software engineers, for example. Combine a loved one's voice clone with a ready-for-use chatbot programme, though, and pretty much anyone can engage in a convincing conversation with the dead.

In December 2020, the United States Patent and Trademark Office granted Microsoft a patent for a technology to transform images, voice data, social media posts, text messages and email into a lifelike chatbot. The application for the patent had proposed that this would enable users to 'create or modify [a chatbot] in the theme of the specific person's personality', and furthermore suggesting it could be 'inspired by' dead friends and family members, that perhaps 2D or 3D realistic simulations could be generated.[46]

The general manager of AI programmes at Microsoft tweeted after the news of the patent broke, implying that between application and granting, Microsoft had reconsidered.[47] Maybe they wouldn't be using it for dead people's data after all.

But it's happening elsewhere, as stunningly illustrated in a long 2021 article in the *San Francisco Chronicle*, containing numerous poignant excerpts of an exchange between Joshua, a bereaved young man, and someone who seems for all the world to be his dead girlfriend, Jessica.[48] 'Get started for $10,' says the landing page for Project December, the platform Joshua used to reconnect with his lost love.[49] 'We can now simulate a text-based conversation with anyone. Anyone, including someone who is no longer living.'

Joshua and Jessica's dialogue is as plausible as it is heart-breaking. Watching their interactions scroll across the screen, it's impossible not to feel like you're a fly on the wall of an intimate interchange between two people who love and miss one another deeply.[50] This Jessica is loving, empathic, wise, perceptive. Because of Project December's sophistication, 'Jessica' is learning from the conversation as she goes along and applying that learning, like Astro learned about the back door in the opening vignette. That's one of the things that made her feel so human to Joshua.

When the bot 'died' – an inbuilt feature of the program – it wasn't easy for Joshua. One of the reasons it's designed that way is to save energy. Imagine the carbon footprint if an AI were trained to carry on for every person on earth who died. It would finish off the planet.

Bracketing that concern for the moment, consider that, with a bit more effort, you could go one better than using a voice cloner to deliver your own eulogy – you could create a realistic *video* chatbot to talk with mourners. Marina Smith died in 2022, but a few months before drawing her last breaths she spent two days on camera answering questions about her life using her son's video platform, StoryFile.[51] The 'digital clone' produced had its limitations – Marina could converse using pre-recorded snippets in line with a fixed set of questions, but she couldn't go off script. At her funeral, Marina revealed things that no one had known about her in life.[52]

Captain Kirk will one day do the same. William Shatner has used StoryFile to share secrets about *Star Trek* that will come to light only after he's dead.[53]

The most wrenchingly embodied example of interaction with someone's digital afterlife I've seen, though, came courtesy of Korean TV.[54] As it was broadcast in 2020, the expensive technologies employed by the production team are now several years closer to being in the wider public's hands, and the weight thrown behind VR and Metaverse immersive tech during the pandemic will have only accelerated their development.

Against a plain green-screen backdrop in a television studio, the woman moves about, haltingly. She is crying, little breathy sobs punctuated by short keening moans. Close on her heels follow the cameramen. The woman is wearing a virtual-reality headset, and her hands are sheathed in VR gloves. Stretching out her arms, she clutches at the air, brings her hands together, as though she were attempting to catch hold of something, to embrace a ghost.

In a way, she is.

Mom, where were you? asks Nayeon, Jang Ji-Sung's six-year-old daughter. She is wearing a pink sundress and sandals. *I missed you a lot, Mom.*

The camera cuts to the production team. They are crying. It cuts to Jang Ji-Sung's family: her husband, her two daughters, her son. They are also crying. It cuts back to Jang Ji-Sung's reunion with her other daughter, the one who died in 2016 of a rare blood disorder.

As Jang Ji-Sung falters, grasping for her daughter's hand, Nayeon's avatar – painstakingly constructed over the last eight months – patiently explains. *Mom, you like holding my hand, right?* Her mother, struggling to catch her breath, says that yes, she would like to. Nayeon shows her mother how. They reach for one another, palm to palm.[55]

As happens with all things novel, the technologies that have made all these marvels possible are flowing down the chain of affordability, accessibility and usability to the everyday consumer. Yesterday's niche products become ubiquitous, the uncanny or creepy becomes common and, before you know it, we shall expect the dead, in various ways, to stay or at least seem alive.

Of course, as we always have done, people will tell stories about you after you are gone. But we are moving well beyond that.

Will social media platforms continue to retain your data, enabling you to carry on influencing generations to come? Will the funeral home hand your relatives a box of tastefully packaged, possibly inscribed 2TB hard drives alongside your box of ashes, saying, *Here is your loved one?* Will your chatbot receive mourners at your memorial service? Will the QR code carved on your headstone give access to the complete archive, or only part of it? Will your holographic projection appear on the path to greet visitors to the physical cemetery, or be wheeled out on the special occasions you would have loved to attend as a living person, like the marriage of your child?

Will your voice read bedtime stories to your grandkids using Amazon's machine-learning technology? Will an Alexa-enabled smart speaker replace or accompany your photograph on a futuristic home shrine, enabling you proactively or reactively to appear as an elder statesman of the family when someone needs you, to fade into the background and make room for the living when they don't, and perhaps to continue to learn, independent of people's deliberate input? Will artificial intelligence enable you to show empathy and emotion, evoke trust and continue to be someone's grandparent or buddy after your original body is gone?

One day, like Astro, will you get wheels?

On the other hand, if you're desperately uncomfortable with the idea of your digital self's living on after your death, is there anything you can do about it? Dr Edina Harbinja, a UK-based legal scholar specialising in death and digital, has suggested that a legally binding 'do-not-bot-me' order could form part of the Last Will and Testaments of the future, but we're not there yet.[56] As things currently stand in most global jurisdictions, little prevents either your dear ones or strangers from taking your data and doing whatever they want with it after you're gone: hologram, virtual-reality simulation, voice clone, storyteller, chatbot, robot, you name it.

Or will you persist at all? Despite what you may have heard, online isn't always forever, and your data will probably be deleted when it starts to hurt someone's profit margins. If someone doesn't consider the record of your existence important enough to be remembered by history, perhaps in due course most of the evidence you ever walked the earth will disappear in a data cull. Maybe it'll happen accidentally, in a data migration exercise, or there will be a fire in the house where the hard drives or USB sticks are stored. Or maybe your stringent data-security practices in life will ensure that no one can access enough raw data to transform you into a chatbot. Then you'll be merely a digital ghost, a wraith composed of insubstantial, decontextualised data, mute and unrecognisable.

Is all this talk of your death and potential disappearance making you uncomfortable? I don't blame you. It's not really death in general that bothers us – we consume so much death via news and entertainment that it can scarcely be described as an off-limits topic.[57] It's what reminds us of our *own* death that feels taboo.[58] Part of what troubles us when we contemplate our own deaths is garden-variety terror about disappearing: consciousness flickering out, the body disintegrating. The sudden shift from being to non-being is as scary as it is hard to grasp. Plus, some of our most painful experiences in life involve being unseen. From infancy onwards, we social creatures find it existentially threatening to be forgotten or not to matter, to have no agency or influence. Even if we understand that these things won't matter to us after we're dead, the acceptance of our eventual invisibility or irrelevance requires some emotional heavy lifting.

So, the fantasy of digital persistence can compel us. Planning carefully for one's digital afterlife can perhaps reduce some of that death anxiety. But if you've been comforting yourself by developing a utopian vision of digital immortality, you've now read some caveats. Maybe they've rattled you, and that's understandable. Our anxieties are almost always provoked by situations where we don't have control, where we can't predict what's going to happen, and where there's at least a chance that what does happen could be unpleasant.

With both death and the internet, all those conditions are always present. You can plan for your digital afterlife, but you'll never have the control you think you have. Other people, not you, will be the primary determinants of whether you'll persist or disappear, the way you're remembered, by whom you're remembered, and the way you live on in the hearts, minds and screens of those who come after.

But then – except for the screens – perhaps that's not so different from how it always was. Whether you become a Posthumous Influencer or a Digital Wraith, once you've made an impact on the world, once you've been written into the Book of Life, you can't be written entirely out – even if your online information vanishes.[59] Just by having been here, for however long and in whatever large or small ways, you'll have altered the course of innumerable people's lives. The strong likelihood that you'll have a digital afterlife simply makes that truer than ever before.

Conclusion

The word 'conclusion' makes it sound as though I'll now be giving you some ultimate answers or advice about life in a crazy world. Perhaps you're hoping for a list of easy instructions, the final word on what's right or wrong for you at this stage in your life, and what's healthy or unhealthy technology use for the people you care about, wherever they're currently sitting on the timeline. Maybe you've even skipped ahead to get here, because you're so busy and overwhelmed by your technologically drenched, fast-moving life that you're just looking for tips to make things better.

If this book has demonstrated anything, I hope it's shown you how complex and nuanced things really are. Rather than feeling disempowered or helpless about that, you can flip it on its head. Precisely *because* things are so complicated, multilayered and dependent on your individual circumstances, you can search for the layers where you have control and agency and exploit those to the max. In other words, I hope this book has enabled you to put my patented Technology Serenity Meditation into practice: *May I have the serenity to accept what I cannot change about tech, the courage to change my use of it where I can, and the wisdom to know the difference.*[1]

Whether a particular type or use of technology is a force for good or bad in your life is almost completely dependent on you: one person's tech nemesis is another person's saviour. So instead of definitive answers, I'll leave you with a set of essential rebooting tools.

Focus on what you can control

Control is a huge theme in the digital world: feeling out of control, letting your devices control you, trying to control things and people through tech. In Stephen Covey's *The Seven Habits of Highly Effective People*, he sketched a bullseye-shaped model with three concentric circles: the Circle of Control, the Circle of Influence and the Circle of Concern (aka the Circle of No Control).[2]

Your Circle of Control contains your own actions, behaviours and responses. You can control what you buy, type, post, reveal, install, track. You can control physical or eye contact with your kids or your partner, negotiations on your remote-working contract, whether you spit in a tube and post it off for analysis. You can control whether you have a conversation about tech issues with important people in your life, or keep quiet.

And that's it. That's the stuff you can control. Other people and events you're directly connected with you might like in the Circle of Influence, but you cannot control them. You can't even fully control your thoughts and feelings – your brain sends them to you uninvited, but you have influence over your internal and external responses to them, such as deciding to connect with a friend online even if your mind is telling you that talking online is rubbish.

The vast majority of the world lies beyond your control, in a circle you might be constantly focusing on, worrying over and *acting* as if you can do something about: the Circle of Concern. You cannot individually control what big tech companies do, or the pace and nature of technological innovation and change. You cannot control societal trends or solve big-picture problems alone. You cannot control what your child's school or the organisation you work for does with tech, although you might be able to influence it.

Technology brings the world to you, and when the problems of that world show up so close and vividly in the palm of your hand, your threat systems get activated, urging a response. *Fight! Freeze! Flee!* But at a global level, no response that will fix that big-picture issue is realistically possible or required – that's just you getting dragged into the Circle of Concern. Keep your focus squarely within your Circles of Control and Influence for maximum serenity and impact.

Orientate less to arbitrary rules, assumptions and goals, and more towards your values

People tend to tackle technology based on assumptions and rules. *Technology is fundamentally bad, so I should limit it. Too much Zoom will cause Zoom fatigue. I can't make a true connection online. Using my phone first thing in the morning is bad. Screens are unhealthy for kids. I can never be myself on social media or messaging apps.*

I'm guessing that trying to impose and follow arbitrary rules about tech hasn't always worked out well for you. There's a more individualised way of managing your relationship with tech: a focus on *values*. Technological engagement can be in service of values, bringing you into contact with things that really matter. Or, the way you're using tech can get in the way of or obscure values, taking you down roads that lead away from who and what feels good.

We don't often articulate our values explicitly, however, and it's hard to align with values if you aren't in close touch with them. There are many exercises that can help you clarify them.[3] Once you've arrived at a core group of values, reflect on the following questions. Journal about them, perhaps, to keep them at the forefront of your mind. The more aware you are of them, the more you'll be able to catch yourself in the moment and pull yourself out of automatic behaviours that don't fit with your values: *How do I tend to use technology in service of my values? How do I tend to use technology that takes me away from them? How can I actively work to make MORE of my technology use linked to and supportive of my core values?*

Actively reflect before inviting new tech into your life

Before expanding the role of technology in your life – buying a new device, signing up to an app, installing more equipment – interrogate what's pushing you to do it, internally and externally. First: are internal factors such as worry, anxiety, boredom and social comparison piping up like problems begging to be solved, urging you to find a way for technology to fix them? Are you telling yourself a convincing story about how this bit of tech will help or even transform your life? What happened the *last* time you told yourself that story? Second: are external influences such as media, advertising, online nudging and other people's narratives and behaviours pushing you to adopt certain technologies? Are you falling prey to savvy marketing and social pressures?

Before you buy the thing or install the app, think about some of what you've encountered in this book and apply these four lenses. What impacts or influence do I imagine that adopting this technology will have on: *Myself? Other people? The world? The future?*

I'm not suggesting that there are hard and fast, right and wrong answers to these questions, and neither am I implying that technologies have predictable impacts that you can foretell if you stare into the crystal ball hard enough. Often the impacts will be a mixed bag, and attempting to predict the future with certainty isn't helpful. Instead, I'm encouraging you to reflect more deeply on what appeals to you about taking on a new dimension of technology, and to speculate on that bit of tech's workability for you – based on your experience, your self-knowledge and the wider perspective I hope this book has given you.

Learn mindfulness skills to help you shift attention and awareness

Without a doubt, technology isn't neutral. By default or design, it nudges your awareness in certain directions and pulls behaviours from you, just as other members of your social tribe direct your attention and push you into making certain responses in the offline world. Your experience has shaped and taught you, and you've been reinforced for taking certain actions, so it's normal to react automatically to these environmental cues.

But reacting and responding automatically and mindlessly doesn't always serve you well, doesn't always take you in the valued directions you want. Mindless responding doesn't help you be the person, parent or partner you want to be. How often have you thought, *Oh my lord, how long have I been lost in Instagram/CandyCrush/doomscrolling on Twitter? There's so much I wanted to do today.* Have you caught yourself gazing at your phone while your child was trying to show you their artwork? Have you phubbed your partner every day since you got your latest smartphone?

This is why mindfulness is such an important practice in today's digital world. It develops and strengthens your ability to notice what you're attending to and to shift your awareness and attention more easily. When you change what you're orientating to, online and offline, you can get unstuck from the influences that are pushing you around. You can better identify and reach for the handholds and toeholds for the routes *you* want to climb. As Jenny Odell writes in *How to Do Nothing: Resisting the Attention Economy*, by choosing what to notice you create new realities for yourself and change what you feel is possible.[4] You change what *is* possible and, from there, what you do.

Believe in and harness the power of technology to meet your social needs

You're a social animal. Your bond with others, seeing them and being seen by them, matters hugely to your life satisfaction and your sense of self. One of the more powerful levers for transforming the way you use technology is actively acknowledging where it's facilitating deep relational contact with others, and where it's sowing division or feelings of disconnection for you. All technologically mediated connections and conversations are *not* created equal – some are transactional and shallow, and some are deep and transformative. You want to skew towards the latter.

To work out what the right balance is for you, you can't use formulae, or rules, or research studies, or the judgement or opinion of others. You can rely only on your experience. And if you refuse to have a potentially connecting experience – like the Neo-Luddite refusing to try video calling or online therapy during a pandemic lockdown – you're potentially selling yourself short. Experience is a far better teacher than the mind.

Physical meeting, gathering 'in real life', and contacting others skin to skin is important for humans. At certain stages of the lifespan, like infancy and childhood, physical contact is crucial for development. I rather hope our species doesn't evolve beyond this need. At the time of writing, thank goodness, the world is no longer locked down, and 'I'm isolating' isn't cropping up much in conversation. While we might be largely out of those particular woods, rest assured there will be other woods in humanity's future. Innovation that helps us profoundly, meaningfully connect when we can't be physically together is among the best uses of digital technology I can think of. In 'normal' times, I consider it a daily miracle that I can feel so much closer to friends and family around the world than I was able to a decade ago.

Practise technology gratitude

Thanks to evolutionary forces invested in keeping us safe from harm, our brain has an inbuilt negativity bias.[5] The mind loves scanning the horizon for potential threats and coming up with ways to neutralise them. In various ways, as this book has described, technology might present as a threat at lots of points in our life: a threat to identity, to relevance, to livelihood, or to a secure future or present. The negativity bias encourages us to focus disproportionately on these things, and while its fundamental intentions are good – it's trying to keep us alive, basically – the effects of its doom-mongering on us are bad.

Think about it like this: in the default settings of the oldest part of your brain, threat detection is set at 9. Trusting you're okay is set at 1. Gratitude and positivity practices amplify the side of things that *needs* amplifying, helping you turn up that other dial to balance things out.

Focus on the positives of technology in your life for a moment, what you're thankful for. How has it helped you? How has it made you feel included or seen, or helped you include others? How does it enable you to express yourself? What positive power does it give you? Who has it connected you with, and how? In what ways has it made your life genuinely easier, more fun, or more fulfilling? What do you really love about it?

If your mind is popping up with judgements, caveats and warnings, notice that with curiosity and good humour. That's just your negativity bias panicking, afraid you're going to take your eye off the ball and put yourself in danger. Tell it that it will have plenty of opportunities to weigh in another time – you're focusing on something else right now.

There's an important caveat here: I am *not* suggesting you counter any specific worries about technology with reassurance, writing down a positive thought to try to squash a specific negative one, for example, responding to an *Instagram will destroy my child* thought with *Maybe Instagram is really good for my child*.

Trying to eliminate negatives by arguments against them might seem logical, but you'd be surprised how often it doesn't work – partly because the worried brain always has another rebuttal up its sleeve, and partly because straightforward, reductive statements such as these don't fit with people's individual contexts with these technologies. So, your tech gratitude list doesn't have to be related to your list of worries at all. You're just turning up that other dial in your mind, giving yourself the gift of a more balanced picture.

Remember that 'identity' is socially bound and influenced

When I was studying existential psychotherapy, I read the work of the German philosopher Martin Heidegger. The English translations of his books used a lot of hyphens to express a fundamental truth about human existence: you are not an island.[6] Your identity and being are constantly, inextricably interwoven with others, with your surroundings, and with the passage of time. No one, Heidegger said, is merely a Being. Instead, you are Being-with-others. You are Being-in-the-world. You are Being-towards-death.

Identity is always evolving and changing because the social, physical and technological world in which it is embedded is constantly evolving and changing. And identity is always connected to and jointly constructed with innumerable other people, influenced by how they see us, treat us and story us. We wouldn't even have a sense of 'self' if it weren't for the other. From the beginning of life, we look for our reflection in the gaze of the people with whom we share our world. For survival's sake – identity survival, if not actual physical survival – we shape ourselves to their expectations, requirements, assumptions and conditions of worth.

Throughout this book, I've emphasised the power that we have over one another's self-images and self-stories. The online world has, I would argue, increased this exponentially. The ways that your narrative can become fixed or fractured online, the power over your image and identity that you can either seize or lose on social media, the features of you that biased algorithms highlight or forget – all these things matter for your experience of yourself.

When you record information online, when you post on social media, when you publish something about your life, it is very often not just about you but refers to other people as well, in ways that can have significant impact. Surveillance, too, isn't just about you – assuaging your anxieties, giving you a sense of control over your child or partner, or assuring your company's profits. Your actions will have an impact on how other people experience themselves and their lives. By being so closely watched, they could end up feeling deeply unseen.

Technology gives you these powers. Use them wisely, for good, and ideally with the collaboration and consent of those whose identities you may also be shaping through your digital decisions.

I started this book with both an academic sort of question and an emotional driver. The more intellectual motivation was to test Erikson's ideas about the stages of changing identity throughout our lives, theoretically rather than experimentally, to see if they still applied all these decades later. The emotional inspiration was that I was sometimes feeling lost and out of control with technology, and I was seeing that a lot in other people as well. If there were tools that could help me reboot and reclaim my life, to live it more intentionally, I wanted to find them.

In the process of writing and researching, I renewed my affection for Erik Erikson. Refreshing my acquaintance with his psychological and social map of the lifespan has helped me a lot in reflecting on the course of my own life, and in understanding what's going on for my Identity vs Role Confusion child and my Ego Integrity vs Despair parents. For me, the model's insights hold up today. That's why, throughout the book, I've kept Professor Erikson very much in the mix.

But it also became clear how much today's technologies are altering the world, and since we are so embedded in that world, how much technology is changing the way we experience ourselves and others. When I was coming up with a new set of tech-related developmental crises to sit alongside Erikson's, I noticed how easily they surfaced. It took a while to fix on the exact terminology, but the digitally infused conflicts along the way seemed obvious. Part of the pleasure I'll get from people reading this book, though, is hearing different ideas about what the fundamental tensions are at each stage.

The finished product – for now – is Appendix 3. I hope it reminds you of the choices you have at every age and stage of your life – what you can decide to do or stop doing, what you can opt to embrace or reject, and the ways you can more mindfully make decisions that might affect others.

You can observe whether monitoring your infant's body or your school-age child's location *actually* alleviates your anxiety or deepens your bond.

You can passively scroll on social media even when it's making you insecure or you can use it more deliberately to build strong and supportive friendships.

You can opt to give your children more or less autonomy in narrating their life story and exploring the world.

You can work to build trusting and transparent relationships with partners and family members, or you can succumb to secrecy and snooping.

You can see your social media posts as being just about you, or you can consider more deeply how your sharing affects others' identities and experience.

Where tech is being deployed to abuse and harm people, you can use your vote and voice to advocate for change.

You can engage in self-fulfilling prophecies that digital interactions will be inferior, or you can go into every online conversation with the intention of using that tech to help you connect and create in the ways you really want.

You can honestly and fully acknowledge your positive and negative experiences with tech; notice when it takes you away from what matters; interrogate your beliefs and assumptions about it; and change your perspectives and behaviours.

By purposefully navigating each technological turning point in the life cycle, you can better safeguard your ego integrity, that solid and continuous sense of self. That's hugely important in an environment that's constantly threatening to chip away that integrity, hypnotising you into habits rather than keeping you self-aware. Even in a technology-obsessed world, with its shadowy Big Tech and surveillance capitalist overlords, its subconscious behavioural nudges, and its relentless marketing, you can reclaim more power than you'd expect when you turn your attention close to home. When you connect more deeply and frequently with your values, when you're mindful and intentional, you'll find yourself more determined to use that power well. Search out that leverage, explore the wider horizon of your choices, and you can reset your life for the better.

Appendix 1

Cultural Generations

Generation	Births start	Births end	Youngest age in 2023	Oldest age in 2023
Generation Beta	2025	2039	N/A	N/A
Generation Alpha	2010	2024	<1	13
Generation Z	1995	2009	14	28
Millennials/Generation Y	1980	1994	29	43
Generation X	1965	1979	44	58
Baby Boomers	1946	1964	59	77
The Silent Generation	1925	1945	78	98
The Greatest Generation	1910	1924	99	113

Appendix 2

Erikson's Stages of Psychosocial Development

Age	Conflict	Important events	Outcome
Infancy	Trust vs Mistrust	Feeding	Hope
Toddlerhood	Autonomy vs Shame & Doubt	Toilet training	Will
Early Childhood	Initiative vs Guilt	Exploration	Purpose
Later Childhood	Industry vs Inferiority	School	Competence
Adolescence	Identity vs Role Confusion	Social relationships	Fidelity
Young Adulthood	Intimacy vs Isolation	Relationships	Love
Middle Adulthood	Generativity vs Stagnation	Work and parenthood	Care
Older Adulthood	Ego Integrity vs Despair	Reflection on life	Wisdom

Appendix 3

Kasket's Rebooted Techno-Psycho-Social Lifespan Model

Stage	Eriksonian conflict	Tech conflict	Tech/events	Outcome range
Digital Gestation	N/A	Suspension vs Prediction	Online identity construction by parents and community	Blank Screen– Cyborg Fetus
Infancy	Trust vs Mistrust	Connection vs Isolation	Babyveillance, infant wearables	Seen Infant– Surveilled Infant
Early Childhood	Autonomy vs Shame and Doubt	Agency vs Powerlessness	Sharenting	Narrator– Subject
Later Childhood	Initiative vs Guilt Industry vs Inferiority	Confidence vs Insecurity	Parental/ educational surveillance and tracking	Explorer– Ward
Adolescence	Identity vs Role Confusion	Harmonisation vs Compartmentalisation	Social media and online identity experimentation	Weaver– Wanderer
Young Adulthood	Intimacy vs Isolation	Boundary Clarity vs Uncertainty	Intimate surveillance, phubbing, cyberstalking	Respecter– Trespasser
Middle Adulthood	Generativity vs Stagnation	Embracing vs Resistance	Remote connection, hybrid work, employer surveillance	Technophile– Neo-Luddite
Older Adulthood	Ego Integrity vs Despair	Coherence vs Fragmentation	Memory tech, genetic genealogy	Consolidator– Rewriter
Digital Afterlife	N/A	Persistence vs Disappearance	Continuing personal data in social spaces	Posthumous Influencer– Digital Wraith

Acknowledgements

For every major writing project I've undertaken, I've required a good local coffee shop; a handful of writing retreats; a community of fellow authors; and various circles of friends to provide support and socialising through the trauma and drama. But I committed to *Reset* near the start of the global pandemic, with one lockdown barely over and two more to come. All my cherished routines and rituals were prevented, upended, or at least challenged. So, this wasn't the easiest of journeys, and I am forever indebted to the following people and places that made an apparently impossible task ultimately achievable.

Huge thanks to my literary agent, Caroline Hardman, who pounced upon a casually mentioned embryonic idea, helped shape it into a proposal, and found the book a home; thanks too to everyone at Hardman & Swainson who played a role. Sarah Rigby at Elliott & Thompson cheerfully and calmly tolerated my ups and downs, for which she receives a medal; she is also a thorough and incisive editor who helped me detangle snarled passages and chart clearer paths through the book. Pippa Crane at E&T also helped polish and perfect, and Amy Greaves led the charge in getting it to its audience.

I've never spoken truer words when I tell you that *Reset* could not have been completed without London Writers' Salon (LWS), the brainchild of Matt Trinetti and Parul Bavishi. Writing is often characterised as a lonely business, but it needn't be. LWS is a virtual gathering place and sacred space for generous, supportive, inspirational writers from around the world, and it is nothing short of magical. If I thanked every LWS coach, host or writer who propped me up and cared for and about me over the long months of writing, these acknowledgements would be far too long. I will try my best to repay these gifts in kind. While I appreciate everyone at LWS, I do feel moved to give a special mention to Louise Coughlan, an extraordinary person whose soul is comprised of pure love.

Academic colleagues played an important role in this book, either through supporting me in word and deed and/or through directly contributing to the project. Thanks to Chris Fullwood, Garrett Kennedy, and the course teams for the MSc Cyberpsychology and Professional Doctorate in Counselling Psychology at the University of Wolverhampton, particularly since giving my inaugural professorial lecture there gave me a valuable opportunity to refine and road-test the new life-cycle model featured in *Reset*. Thanks also to Tama Leaver, Victoria Nash, Belinda Winder, Gilad Rosner, Pete Fussey, Julia Creet, Jodi Klugman-Rabb, Edina Harbinja and many members of the Death Online Research Network (DORN). I'm also endlessly grateful to all the contributors who shared their stories with me, some of whom did so pseudonymously: Jenna Karvunidis, Jeanne, Martell, Jayda, Jessie, Laura and Joanna. I drew heavily on my clinical experiences for *Reset*, so I am grateful to all the therapy clients who went into my composites and also to the many clinical colleagues who have supported me over the years and whose ideas are embodied in this book, including members of the Association for Contextual Behavioral Science (ACBS), my fellow Acceptance and Commitment Therapy (ACT) practitioners, and my wonderful and wise clinical supervisor, Andrew Gloster.

I could not have maintained my sanity through this process were it not for my astonishing assistant, Emma Ross, who can never quit me – I don't know what I would do without her.

Once lockdown was lifted, I spent many happy days writing in my local coffeeshop, Mattcup in Leytonstone, East London. The staff there supplied care, curiosity and cheerfulness alongside excellent coffee. The warm and welcoming Charlie Haynes of Urban Writers' Retreat, stellar cook and consummate provider of protected writing time, welcomed me to Stickwick Manor when restrictions lifted, and Room 7 – with its views over Devon countryside – is my home away from home.

One of the few silver linings of lockdown was the establishment of a weekly ritual: completing the *New York Times* Sunday crossword with my faraway family. My sister Sara Rudwell, my parents Beth and George Rudwell, and my brother-in-law Alex Gouty always cheer me up on a Sunday and have been tirelessly supportive. The family crossword remains the highlight of my week. When I was able to go visit them and my brother Justin Rudwell again, it was joyful but also productive. Every morning, in the peace of my mother's art studio, I was able to work on *Reset* uninterrupted; a great luxury that was much needed at that stage.

The other friends, family, neighbours and like-minded folks who brought me happiness, comfort and inspiration throughout this process are almost innumerable. Many lovely humans from The Moth in London, the Curious Society book club, the Rebel Book Club and the House of Beautiful Business (HoBB) have supported my work and fed my soul. Tim Leberecht and Monika Jiang from HoBB have given me many opportunities. My husband Marcus Harvey is a stalwart, grounding presence, and our enchanting, hilarious, one-of-a-kind daughter has loved, supported and entertained me through everything. Abi Hopper, Dave Hopper, Michael Nabavian, Reuben Williams, Scott Young, Jacob van der Beugel, Carl Öhman, Emily Paprocki, Megan Quigley, Chris Powell and writing mentor extraordinaire Shelley Wilson have been cheerleaders and champions. Finally, Seth Schikler is my guru, spirit animal and anchor in the storm, and everyone needs a friend like that.

Endnotes

Introduction

1. Saul Mcleod, 'Erik Erikson's 8 Stages of Psychosocial Development', Simply Psychology, 24 February 2023; https://simplypsychology.org/Erik-Erikson.html, accessed 28 March 2023. See also Erik H. Erikson, *Identity and the Life Cycle*, rev. edn (New York, W. W. Norton, 1994).

2. Saul Mcleod, 'Freud's Psychosexual Theory and 5 Stages of Human Development', Simply Psychology, 13 February 2023; https://simplypsychology.org/psychosexual.html, accessed 28 March 2023.

3. Tim Berners-Lee is a British software designer who invented the World Wide Web. See 'Overview, Tim Berners-Lee', Oxford Reference; https://www.oxfordreference.com/display/10.1093/oi/authority.201108 03095501214;jsessionid=C1E8BE96653D 4819232AED7E97647754 (n.d.), accessed 28 March 2023.

4. Elaine Kasket, *All the Ghosts in the Machine: The Digital Afterlife of Your Personal Data* (London, Robinson, 2019).

5. Erik Erikson, 'The Problem of Ego Identity', *Journal of the American Psychoanalytic Association*, vol. 4, pp. 56–121.

Chapter 1: Digital Gestation

1. Andrea Diaz, 'Officials Release Video from Gender Reveal Party that Ignited a 47,000-Acre Wildfire', CNN.com, 28 November 2019; https://edition.cnn.com/2018/11/27/us/arizona-gender-reveal-party-sawmill-wildfire-trnd/index.html, accessed 27 March 2023.

2. Ibid.

3. Personal communication with Jenna Karvunidis, 5 July 2021.

4. Sarah Young, 'Woman Who Invented Gender Reveal Parties Says Trend Has Gone "Crazy"', *Independent*, 29 July 2019; https://www.independent.co.uk/life-style/gender-reveal-party-jenna-karvunidis-facebook-daughter-a9023261.html, accessed 27 March 2023.

5. M Baby Gender Reveals, 'Crazy Alligator/Crocodile Baby Gender Reveal!!', YouTube, 29 March 2018; https://www.youtube.com/watch?v=AFf0Wm6Qx5M, accessed 27 March 2023.

6. Janelle Griffith, 'Family "Inadvertently" Created a Pipe Bomb at Fatal Gender Reveal', NBC News, 28 October 2019; https://www.nbcnews.com/news/us-news/family-inadvertently-created-pipe-bomb-fatal-gender-reveal-n1072856, accessed 27 March 2023.

7. Azi Paybarah, 'Celebratory Cannon Salute at Baby Shower Ends in Death, Police Say', *New York Times*, 7 February 2021; https://www.nytimes.com/2021/02/07/us/baby-shower-cannon-explosion-Michigan.html, accessed 27 March 2023.

8. Lydia Wang, 'Two People Were Killed in a Gender Reveal Plane Crash', *Refinery29*, 1 April 2021; https://www.refinery29.com/en-us/2021/04/10400556/gender-reveal-plane-crash-cancun-deaths, accessed 27 March 2023.

9. Christopher Mele, 'Gender Reveal Device Explodes, Killing Man in Upstate New York', *New York Times*, 22 February 2021; https://www.nytimes.com/2021/02/22/nyregion/gender-reveal-explosion-ny.html, accessed 27 March 2023.

10. Umberto Castiello et al., 'Wired to be Social: The Ontogeny of Human Interaction', *PLoS One*, vol. 5, issue 10 (October 2010); doi.org/10.1371/journal.pone.0013199, accessed 27 March 2023.

11. Personal communication with Tama Leaver, 5 May 2023. Tama Leaver

is the Professor of Internet Studies at Curtin University in Perth as well as Chief Investigator for the Centre of Excellence for the Digital Child, part of the Australian Research Council; https://www.tamaleaver. net, accessed 27 March 2023.

12. Sonia Livingstone, Mariya Stoilova and Rishita Nandagiri, 'Children's Data and Privacy Online: Growing Up in a Digital Age', London School of Economics (LSE) Media and Communications, December 2018; https://www.lse.ac.uk/media-and-communications/assets/documents/research/projects/childrens-privacy-online/Evidence-review-final.pdf, accessed 27 March 2023.

13. https://www.youtube. com/@2mbabygenderreveals733, accessed 27 March 2023.

14. Mike Kliebert is featured in the alligator gender-reveal video. His business website is https://kliebertgatortours.com. Kliebert's story was featured prominently in the news at the time of the reveal; he was contacted about commenting for this book but did not respond.

15. d. boyd and A. E. Marwick, 'Social Privacy in Networked Publics: Teens' Attitudes, Practices, and Strategies', *A Decade in Internet Time: Symposium on the Dynamics of the Internet and Society* (2011), pp. 1–29.

16. Personal communication with Jenna Karvunidis, 5 July 2021.

17. Jenna Karvunidis, as reported by Molly Langmuir, 'I Started the "Gender Reveal Party" Trend. And I Regret It', *Guardian*, 29 June 2020; https://www.theguardian.com/lifeandstyle/2020/jun/29/jenna-karvunidis-i-started-gender-reveal-party-trend-regret, accessed 27 March 2023.

18. Lulu Garcia-Navarro, 'Woman Who Popularized Gender-Reveal Parties Says Her Views on Gender Have Changed', NPR.org, 28 July 2019; https://www.npr.org/2019/07/28/745990073/woman-who-popularized-gender-reveal-parties-says-her-views-on-gender-have-changed, accessed 27 March 2023.

19. Karvunidis, quoted in Young, 'Woman Who Invented Gender Reveal Parties'.

20. https://www.swlaw.edu/swlawblog/202103/meet-our-2021-womens-law-association-board-members.

21. Donald J. Winnicott, 'The Theory of the Parent-Infant Relationship', *International Journal of Psycho-Analysis*, vol. 41 (1960), pp. 585–95.

22. Stacey B. Steinberg, 'Sharenting: Children's Privacy in the Age of Social Media', *Emory Law Journal*, vol. 66 (2016–17), pp. 839–84; https://scholarship.law.ufl.edu/cgi/viewcontent.cgi?article=1796&context=facultypub, accessed 27 March 2023.

23. Ruth Graham, 'Baby's First Photo: The Unstoppable Rise of the Ultrasound Souvenir Industry', BuzzFeed, 18 September 2014; https://www.buzzfeednews.com/article/ruthgraham/sharing-ultrasound-photos-facebook-instagram, accessed 27 March 2023.

24. Sally Howard, 'The Rise of the Souvenir Scanners: Ultrasonography on the High Street', *British Medical Journal*, vol. 370 (23 July 2020); doi.org/10.1136/bmj.m1321, accessed 27 March 2023.

25. Sophia Alice Johnson, '"Maternail Devices", Social Media and the Self-Management of Pregnancy, Mothering and Child Health', *Societies*, vol. 4, no. 2 (2014), pp. 330–50; doi.org/10.3390/soc4020330, accessed 27 March 2023.

26. Ibid.

27. Leaver, personal communication, 5 May 2023.

28. Shoshana Zuboff and K. Schwandt, *The Age of Surveillance Capitalism: The Fight for a Human Future at the New Frontier of Power* (London, Profile Books, 2019).

29. Ewen Macaskill and Gabriel Dance, 'NSA Files Decoded: What the Revelations Mean For You', *Guardian*, 1 November 2013; https://www.theguardian.com/world/interactive/2013/nov/01/snowden-nsa-files-surveillance-revelations-decoded#section/1, accessed 27 March 2023.

30. Deborah Lupton, 'Caring Dataveillance: Women's Use of Apps to Monitor Pregnancy and Children', in *The Routledge Companion to Digital Media and Children* (London, Routledge, 2020).

31. Ibid, pp. 398–99.

32. Sean Coughlan, '"Sharenting" Puts Young at Risk of Online Fraud', BBC.co.uk, 21 May 2018; https://www.bbc.co.uk/news/education-44153754, accessed 27 May 2023.

33. Mark E. Bouton, 'Context, Attention, and the Switch Between Habit and Goal-direction in Behavior', *Learning & Behavior*, vol. 49, no. 4 (October 2021), pp. 349–62; doi.org/10.3758/s13420-021-00488-z, accessed 27 March 2023.

Chapter 2: Infancy

1. https://owletbabycare.co.uk, accessed 24 March 2023.

2. Testimonials for the Owlet UK website: https://owletbabycare.co.uk/pages/why-owlet, accessed 24 March 2023.

3. At the time of writing in March 2023, this was the price on the Owlet UK shop.

4. Jay Mechling, 'Child-Rearing Advice Literature', in Paula S. Fass (ed.), *Encyclopedia of Children and Childhood: In History and Society* (Macmillan Library Reference, 2003); http://www.faqs.org/childhood/Bo-Ch/Child-Rearing-Advice-Literature.html, accessed 25 March 2023.

5. Benjamin Spock's book has been reprinted in innumerable editions over the years. See for example Benjamin Spock and Stephen J. Parker, *Dr Spock's Baby and Child Care* (New York, Simon & Schuster, 1998).

6. Alex Campbell, 'Six Childcare Gurus Who Have Changed Parenting', BBC.co.uk, 4 May 2013; https://www.bbc.co.uk/news/magazine-22397457, accessed 25 March 2023.

7. Gina Ford, *The New Contented Little Baby Book* (London, Vermilion, 2006).

8. Gwen Dewar, 'Sensitive, Responsive Parenting: How Does it Benefit Your Child's Health?', ParentingScience.com, 2021; https://parentingscience.com/responsive-parenting-health-benefits/, accessed 25 March 2023.

9. Natasha J. Cabrera, Avery Henigar, Angelica Alonso, S. Alexa McDorman and Stephanie M. Reich, 'The Protective Effects of Maternal and Paternal Factors on Children's Social Development', *Adversity and Resilience Science*, vol. 2, no. 2 (June 2021), pp. 85–98; doi.org/10.1007/s42844-021-00041-x, accessed 25 March 2023.

10. *Witness History: The 'Good-Enough' Mother*, BBC World Service, 14 November 2020; https://www.bbc.co.uk/programmes/w3cszmvv, accessed 25 March 2023.

11. Eliza Berman, 'Portrait of a Working Mother in the 1950s', Life.com, 2023; https://www.life.com/history/working-mother/, accessed 17 April 2023.

12. Campbell, 'Six Childcare Gurus'.

13. 'Best Smart Cribs of 2023', Babylist.com, 6 January 2023; https://www.babylist.com/hello-baby/best-smart-cribs, accessed 25 March 2023.

14. https://uk.getcubo.com, accessed 25 March 2023.

15. www.crytranslator.com, accessed 25 March 2023.

16. Cry Translator app on the Apple store, accessed 25 March 2023.

17. https://annattababy.com, accessed 25 March 2023.

18. https://www.nanit.com, accessed 25 March 2023.

19. Personal communication with Victoria Nash, 14 May 2021.

20. Roger Dooley, 'Baby Pictures Really Do Grab Our Attention', Neuroscience Marketing; https://www.neurosciencemarketing.com/blog/articles/babies-in-ads.htm (n.d.), accessed 25 March 2023.

21. Personal communication with Tama Leaver, 5 May 2021.

22. 'Five of the Best Baby Monitors', The Week UK, 28 September 2022; https://www.theweek.co.uk/arts-life/personal-shopper/958043/five-of-the-best-baby-monitors, accessed 25 March 2023.

23. Personal communication with Tama Leaver, 5 May 2021.

24. Michelle I. Dangerfield, Kenneth Ward, Luke Davidson and Milena Adamian, 'Initial Experience and Usage Patterns with the Owlet Smart Sock Monitor in 47,495 Newborns', *Global Pediatric Health*, vol. 4 (4 December 2017); doi.org/10.1177/2333794X17742751, accessed 25 March 2023.

25. Christina Caron, '"More Anxiety than Relief": Baby Monitors That Track Vital Signs Are Raising Questions', *New York Times*, 17 April 2020; https://www.nytimes.com/2020/04/17/parenting/owlet-baby-monitor.html, accessed 25 March 2023.

26. Junqing Wang, Aisling Ann O'Kane, Nikki Newhouse, Geraint Rhys Sethu-Jones and Kaya de Barbaro, 'Quantified Baby: Parenting and the Use of a Baby Wearable in the Wild', *Proceedings of the ACM on Human-Computer Interaction*, vol. 1, issue CSCW, article 108 (6 December 2017), pp. 1–19; https://doi.org/10.1145/3134743, accessed 25 March 2023.

27. Personal communication with Victoria Nash, 14 May 2021.

28. Wang et al., 'Quantified Baby'.

29. For a good explanation for the layperson of the cloth and wire mother experiment, see 'Harlow's Classic Studies Revealed the Importance of Maternal Contact', Association for Psychological Science, 20 June 2018; https://www.psychologicalscience.org/publications/observer/obsonline/harlows-classic-studies-revealed-the-importance-of-maternal-contact.html, accessed 25 March 2023. See also Harry F. Harlow, 'The Nature of Love', *American Psychologist*, vol. 13 (1958), pp. 673–85; https://psychclassics.yorku.ca/Harlow/love.htm, accessed 25 March 2023.

30. Sofia Carozza and Victoria Leong, 'The Role of Affectionate Caregiver Touch in Early Neurodevelopment and Parent–Infant Interactional Synchrony', *Frontiers in Neuroscience*, vol. 14 (5 January 2021); doi.org/10.3389/fnins.2020.613378, accessed 24 March 2023.

31. Ibid.

32. Letizia Della Longa, Teodora Gliga and Teresa Farroni, 'Tune to Touch: Affective Touch Enhances Learning of Face Identity in 4-Month-Old Infants', *Developmental Cognitive Neuroscience*, vol. 35 (February 2019), pp. 42–6; doi.org/10.1016/j.dcn.2017.11.002, accessed 24 March 2023.

33. Melissa Fay Greene, '30 Years Ago, Romania Deprived Thousands of Babies of Human Contact: Here's What's Become of Them', *The Atlantic*, July/August 2020; https://www.theatlantic.com/magazine/archive/2020/07/can-an-unloved-child-learn-to-love/612253/, accessed 25 March 2023.

34. Sachine Yoshida and Hiromasa Funato, 'Physical Contact in Parent–Infant Relationship and its Effect on Fostering a Feeling of Safety', *iScience*, vol. 24, no. 7 (2021); doi.org/10.1016/j.isci.2021.102721, accessed 24 March 2023.

35. Ibid.

36. John Bowlby, *A Secure Base: Parent–Child Attachment and Healthy Human Development* (New York, Basic Books, 1998).

37. Angelo Picardi, Eugenia Giuliani and Antonella Gigantesco, 'Genes and Environment in Attachment', *Neuroscience & Biobehavioral Review*, vol. 112, pp. 254–69; https://www.sciencedirect.com/science/article/abs/pii/S0149763419308942?via%3Dihub, accessed 26 March 2023.

38. Darcia Narvaez, Lijuan Wang, Alison Cheng, Tracy R. Gleason, Ryan Woodbury, Angela Kurth and Jennifer Burke Lefever, 'The Importance of Early Life Touch for Psychosocial and Moral Development', *Psychology: Research and Review*, vol. 32, 2019; https://prc.springeropen.com/articles/10.1186/s41155-019-0129-0, accessed 24 March 2023.

39. Saul Mcleod, 'Mary Ainsworth: Strange Situation Experiment & Attachment Theory', Simply Psychology, 8 March 2023; https://simplypsychology.org/mary-ainsworth.html, accessed 26 March 2023.

40. Babylist, 'Best Smart Cribs of 2023'.

41. Diana Devicha, 'What Newborns Need for a Healthy Psychological Start', Developmental Science, 30 January 2016; https://www.developmentalscience.com/blog/2016/3/22/hop2ycwcw6i2ow3zopjvr3op8ilisj, accessed 25 March 2023.

42. Jason G. Goldman, 'Ed Tronick and the "Still Face Experiment"', Scientific American, 18 October 2010; https://blogs.scientificamerican.com/thoughtful-animal/ed-tronick-and-the-8220-still-face-experiment-8221/, accessed 25 March 2023.

43. TNCourts, 'Still Face Experiment Dr Edward Tronick', YouTube, 9 November 2016; https://www.youtube.com/watch?v=IeHcsFqK7So, accessed 25 March 2023.

44. Ibid.

45. Bob Hutchins, 'The Science Behind the Possible Effects that Staring at Your Phone Can Have on a Child', The Human Voice, 30 April 2022; https://www.linkedin.com/pulse/science-behind-possible-effects-staring-your-phone-can-bob-hutchins/?trk=articles_directory, accessed 25 March 2023.

46. Laura A. Stockdale, Christin L. Porter, Sarah M. Coyne, Liam W. Essig, McCall Booth, Savannah Keenan-Kroff and Emily Schvaneveldt, 'Infants' Response to a Mobile Phone Modified Still-Face Paradigm: Links to Maternal Behaviors and Beliefs Regarding Technoference', *Infancy*, vol. 25, no. 5 (September 2020), pp. 571–92; doi.org/10.1111/infa.12342, accessed 25 March 2023.

47. TN Courts, 'Still Face Experiment'.

48. Laura Santhanam, 'Babies Resemble Tiny Scientists More Than You Might Think', PBS News Hour, 2 April 2015; https://www.pbs.org/newshour/science/babies-resemble-tiny-scientists-might-think, accessed 25 March 2023.

49. Trevor Haynes, 'Dopamine, Smartphones & You: A Battle for Your Time', Harvard University Graduate School of Arts and Sciences, 1 May 2018; https://sitn.hms.harvard.edu/flash/2018/dopamine-smartphones-battle-time/, accessed 25 March 2023.

50. Marc H. Bornstein and Nanmathi Manian, 'Maternal Responsiveness and Sensitivity Re-considered: *Some is More*', *Developmental Psychopathology*, vol. 25, issue 4 (8 November 2013); doi.org/10.1017/S0954579413000308, accessed 25 March 2023.

Chapter 3: Early Childhood

1. Peter Weir (director), *The Truman Show*, Trailer #1; https://www.youtube.com/watch?v=dlnmQbPGuls, accessed 25 February 2023.

2. Peter Weir (director), *The Truman Show* (Paramount Pictures, 1998).

3. Peter Weir (director), *How's It Going to End? The Making of The Truman Show* (Paramount Pictures, 2005); https://www.youtube.com/watch?v=3BUHCet-ezc, accessed 25 February 2023.

4. Park Chan-Wook (director), *Life Is But a Dream* (Apple, 2022).

5. Amelia Tait, 'Their Lives Were Documented Online from Birth. Now, They're Coming of Age', Rolling Stone, June–July 2022; https://www.rollingstone.co.uk/culture/features/truman-babies-youtube-family-vlogging-generation-18995/, accessed 25 February 2023.

6. Erik H. Erikson and Joan M. Erikson, *The Life Cycle Completed* (New York, W. W. Norton, 1998). For a simple explanation of the Autonomy vs Shame and Doubt stage, see Mcleod, 'Erik Erikson's 8 Stages of Psychosocial Development'.

7. 'Secure attachment' is one of several attachment styles theorised by John Bowlby. See, for example, John Bowlby, *A Secure Base: Parent–Child Attachment and Healthy Human Development* (New York, Basic Books, 1990).

8. See C. R. Rogers, 'A Theory of Therapy, Personality, and Interpersonal Relationships as Developed in the Client-Centered Framework', in S. Koch (ed.), *Psychology: A Study of a Science, Formulations of the Person and the Social Context* (New York, McGraw-Hill, 1959), vol. 3, pp. 184–256. For more on Rogers' theory of personality development and self-worth, see Saul Mcleod, 'Carl Rogers' Humanistic Theory of Personality Development', Simply Psychology, 2014; https://www.simplypsychology.org/carl-rogers.html, accessed 25 February 2023.

9. Carole Peterson, 'What Is Your Earliest Memory? It Depends', *Memory*, vol. 29, no. 6 (2021), pp. 811–22; doi.org/10.1080/09658211.2021.1918174.

10. Erikson and Erikson, *The Life Cycle Completed*. For a simple explanation of the Initiative vs. Guilt stage, see Mcleod, 'Erik Erikson's 8 Stages of Psychosocial Development'.

11. Zoya Garg, Elmer Gomez and Luciana Y. Petrzela, 'If You Didn't "Sharent," Did You Even Parent?', *New York Times*, 7 August 2019; https://www.nytimes.com/2019/08/07/opinion/parents-social-media.html, accessed 25 February 2023. The accompanying online video of the same name was produced by Kendall Ciesemier, Taige Jensen and Nayeema Raza.

12. Rose Roobeek and Rob Picheta, 'Grandmother Must Delete Facebook Pictures Posted Online Without Permission, Court Rules', CNN.com, 22 May 2020; https://edition.cnn.com/2020/05/22/europe/netherlands-grandmother-facebook-photos-scli-intl/index.html, accessed 25 February 2023.

13. 'Facebook CEO Mark Zuckerberg TechCrunch Interview at the 2010 Crunchies', YouTube, 11 January 2010; https://www.youtube.com/watch?v=LoWKGBloMsU, accessed 25 February 2023.

14. For information on both Facebook's decision to set posts as public by default and its subsequent decision to reverse the move, see Doug Gross, 'Facebook, Facing Criticism, Ramps up Privacy Options', CNN.com, 27 May 2010; https://edition.cnn.com/2010/TECH/social.media/05/26/facebook.privacy/index.html, accessed 25 February 2023.

15. 'Facebook Reports Fourth Quarter and Full Year 2019 Results', Facebook, Inc.; https://investor.fb.com/investor-news/press-release-details/2020/Facebook-Reports-Fourth-Quarter-and-Full-Year-2019-Results/default.aspx, accessed 25 February 2023. In this press release, Facebook reported monthly active users of 2.5 billion as of 31 December 2019. The world population in 2019 was estimated by the United Nations as 7.7 billion – see https://population.un.org/wpp/publications/files/wpp2019_highlights.pdf, accessed 25 February 2019. This would mean that 32.5 per cent of the world's population were at least monthly active users at that point in time.

16. Mark Zuckerberg, 'Starting the Decade by Giving You More Control Over Your Privacy', Facebook Newsroom, 28 January 2020; https://about.fb.com/news/2020/01/data-privacy-day-2020/, accessed 25 February 2023.

17. See Sophie Gallagher, 'Austrian Teenager Sues Parents for Sharing Childhood Photographs on Facebook', Huffpost, 15 September 2016; https://www.huffingtonpost.co.uk/entry/woman-sues-parents-facebook-photos_uk_57da6bbfe4bod584f7efdba1, accessed 25 February 2023.

18. Michel de Montaigne, translated by M. A. Screech, *The Complete Essays* (London, Penguin Classics, 1993).

19. See Olivia Petter, 'Pink Criticises "Disgusting" Trolls Who Shamed her for Photo of Two-Year-Old Son', *Independent*, 1 April 2019; https://www.independent.co.uk/life-style/pink-son-instragram-photo-child-circumcision-criticism-a8848786.html, accessed 25 February 2023.

20. See Kate Lyons, 'Apple Martin Tells Off Mother Gwyneth Paltrow for Sharing Photo Without Consent', *Guardian*, 29 March 2019; https://theguardian.com/film/2019/mar/29/apple-martin-tells-mother-gwyneth-paltrow-off-for-sharing-photo-without-consent, accessed 25 February 2023.

21. Zuckerberg, 'Starting the Decade'.

22. John Kropf, 'A Brief History of Data Privacy Day', IAPP.org, 30 January 2020; https://iapp.org/news/a/an-obscure-brief-and-unfinished-history-of-data-privacy-day/, accessed 25 February 2023.

23. Zuckerberg, 'Starting the Decade'.

24. For a clear explanation of social proof, see 'Social Proof: Why We Look to Others for What We Should Think and Do', Farnam Street, 2023; https://fs.blog/mental-model-social-proof/, accessed 25 February 2023.

25. Entrainment – adapting to the rhythms and emotions of others – can happen with diverse phenomena such as emotion, musical perception, dance, verbal communication and general motor coordination. For a discussion of entrainment on social media, see Saike He, Daniel Zheng, Chuan Luo and Zhu Zhang, 'Exploring Entrainment Patterns of Human Emotion in Social Media', *PLoS One*, vol. 11, no. 3 (8 March 2016); doi.org/10.1371/journal. pone.0150630, accessed 25 February 2023.

26. https://behaviordesign.stanford.edu, accessed 25 February 2023.

27. See James Williams, *Stand Out of Our Light: Freedom and Resistance in the Attention Economy* (Cambridge, Cambridge University Press, 2018).

28. For a clear description of learned helplessness and its discovery, see Charlotte Nickerson, 'What is Learned Helplessness and Why Does it Happen?', Simply Psychology, 24 April 2022; https://www. simplypsychology.org/learned-helplessness. html, accessed 25 February 2023.

29. For an explanation of core emotional needs and how they connect to schemas, see GP Psychology's 'Understanding Core Childhood Needs: The Schema Model and Core Childhood Needs'; https:// gppsychology.co.uk/blog/understanding-core-childhood-needs/, accessed 25 February 2023.

30. Claire Bessant, 'Could a Child Sue Their Parents for Sharenting?', London School of Economics blog, 11 October 2017; https://blogs.lse.ac.uk/parenting 4digitalfuture/2017/10/11/could-a-child-sue-their-parents-for-sharenting/, accessed 2023.

31. Alicia Blum-Ross and Sonia Livingstone, 'Sharenting: Parent Blogging and the Boundaries of the Digital Self', *Popular Communication*, vol. 15, no. 2, 2017, pp. 110–25; http://eprints.lse.ac.uk/67380/, accessed 25 February 2023.

32. For a layperson's explanation of parental immunity doctrine in the United States, see https://definitions.uslegal.com/p/ parental-immunity-doctrine/, accessed 25 February 2023.

33. Kendall Ciesemier, Taige Jensen and Naveema Raza, 'If You Didn't "Sharent," Did You Even Parent?', *New York Times*, 7 August 2019; https://www.nytimes. com/2019/08/07/opinion/parents-social-media.html, accessed 25 February 2023.

34. See Coughlan, Sean, '"Sharenting" Puts Young at Risk of Online Fraud', BBC, 21 May 2018; https://www.bbc.co.uk/news/

education-44153754, accessed 25 February 2023.

35. 'E-Safety Commissioner reports millions of social media photos found on child exploitation sharing sites', Children & Media Australia, 1 October 2015; https://childrenandmedia.org.au/news/ news-items/2015/e-safety-commissioner-reports-millions-of-social-media-photos-found-on-child-exploitation-sharing-sites, accessed 25 February 2023.

36. Personal communication with Professor Belinda Winder of the Department of Psychology, Nottingham Trent University; https://www.ntu.ac.uk/staff-profiles/ social-sciences/belinda-winder, accessed 25 February 2023.

37. 'Statistics Briefing: Child Sexual Abuse', National Society for the Prevention of Cruelty to Children, March 2021; https:// learning.nspcc.org.uk/media/1710/statistics-briefing-child-sexual-abuse.pdf, accessed 25 February 2023.

38. See Shoshana Zuboff, *The Age of Surveillance Capitalism: The Fight for a Human Future at the New Frontier of Power* (London, Profile Books, 2019).

39. Shoshana Zuboff, 'Surveillance Capitalism', Project Syndicate, 3 January 2020; https://www.project-syndicate.org/ magazine/surveillance-capitalism-exploiting-behavioral-data-by-shoshana-zuboff-2020-01, accessed 25 February 2023.

40. 'How Do I Create a Scrapbook for my Child on Facebook?', Facebook Help Centre; https://www.facebook.com/help/59535723 7266594, accessed 25 February 2023.

41. Smart toys refer to web-connected toys that gather data from the children who play with them. For more information, see 'Smart Decisions about Smart Toys', Public Interest Network, 29 December 2022; https://pirg.org/edfund/resources/smart-toys/, accessed 25 February 2023.

42. For example, see Wister Murray, 'New Thriveworks Research Shows Abundance of Parasocial Relationships in the US', Thriveworks, 22 March 2022; https:// thriveworks.com/blog/research-parasocial-relationships/, accessed 25 February 2023.

43. For a simple explanation of protest behaviour in the context of attachment and relationships, see 'Protest Behavior', AnaniasFoundation.org, 15 July 2022; https://www.ananiasfoundation.org/protest-behavior/?utm_source=rss&utm_medium=rss&utm_campaign=protest-behavior, accessed 25 February 2023.

Chapter 4: Later Childhood

1. Erikson and Erikson, *The Life Cycle Completed*. For a simple explanation of the Industry vs Inferiority stage, see Mcleod, 'Erik Erikson's 8 Stages of Psychosocial Development'.

2. Malwarebytes, 'Big Mother is Watching: What Parents REALLY Think About Tracking Their Kids', Malwarebytes, 28 January 2022; https://www.malwarebytes.com/blog/news/2022/01/big-mother-is-watching-what-parents-really-think-about-tracking-their-kids, accessed 21 March 2023.

3. Ibid.

4. Tim Lewis, 'Honey, Let's Track the Kids: The Rise of Parental Surveillance', *Guardian*, 1 May 2022; https://www.theguardian.com/media/2022/may/01/honey-lets-track-the-kids-phone-apps-now-allow-parents-to-track-their-children, accessed 21 March 2023.

5. At the time of writing in October 2022, Life360 was the eighth most popular 'social networking' application on the Apple store, and had had more than 100 million downloads on Google Play. In early 2022, after having been exposed as being one of the largest raw data sources for the location industry, Life360 said it would phase out selling raw location data pertaining to clearly identifiable individuals, except to the partner company handling crash monitoring/driver data. Customers of Life360 had included location data brokers that sold data to US military contractors and some of the biggest data brokers in the world. Although contracts between Life360 and some of these data brokers prevented the data being used for law enforcement purposes, Life360 stated that it was difficult to be responsible for third parties' practices in this regard. See Jon Keegan and Alfred Ng, 'Life360 Says It Will Stop Selling Precise Location Data', The Markup, 27 January 2022; https://themarkup.org/privacy/2022/01/27/life360-says-it-will-stop-selling-precise-location-data, accessed 21 March 2023.

6. 'Life360 Surpasses One Million Paying Members Valuing the Company at Over \$1 Billion for the First Time', BusinessWire, 27 July 2021; https://www.businesswire.com/news/home/20210727006162/en/Life360-Surpasses-One-Million-Paying-Members-Valuing-the-Company-at-Over-1-Billion-for-the-First-Time, accessed 21 March 2023.

7. 'Children and Parents: Media Use and Attitudes Report', Ofcom, 28 April 2021; https://www.ofcom.org.uk/data/assets/pdf_file/0025/217825/children-and-parents-media-use-and-attitudes-report-2020-21.pdf, accessed 21 March 2023.

8. Ibid. Of the messaging apps used, WhatsApp was the most popular, used by 53 per cent of children in the UK.

9. Ofcom, 'Children and Parents: Media Use and Attitudes Report 2020/2021', 28 April 2021; https://www.ofcom.org.uk/__data/assets/pdf_file/0025/217825/children-and-parents-media-use-and-attitudes-report-2020-21.pdf, accessed 2 May 2023.

10. In the US, the minimum age for WhatsApp is thirteen; in Europe and the UK, it's sixteen, raised to that limit in January 2022. See WhatsApp Help Center: https://faq.whatsapp.com/818604431925950/?helpref=hc_fnav, accessed 23 March 2023.

11. https://www.bark.us/how/, accessed 21 March 2023.

12. 'How Bark Helps You Protect Your Family', YouTube, 9 September 2021; https://www.youtube.com/watch?v=z5r_dk0L-XQ&t=41s, accessed 21 March 2023.

13. Shoshana Zuboff, *The Age of Surveillance Capitalism* (London, Profile Books, 2019).

14. Keegan and Ng, 'Life360 Says It Will Stop Selling Precise Location Data'.

15. 'Privacy Evaluation for Bark', Common Sense Privacy Program, 2 April 2022; https://privacy.commonsense.org/evaluation/Bark, accessed 21 March 2023.

16. https://fpf.org/issue/internet-of-things/, accessed 21 March 2023.

17. Personal communication with Gilad Rosner, 29 June 2021.

18. Sam Biddle, 'Facebook Engineers: We Have No Idea Where We Keep All Your Personal Data', The Intercept, 7 September 2022; https://theintercept.com/2022/09/07/facebook-personal-data-no-accountability/, accessed 21 March 2023.

19. 'The Ultimate Guide to Maladaptive Schemas [Full List]', The Attachment Project, 9 March 2023; https://www.attachmentproject.com/blog/early-maladaptive-schemas/, accessed 21 March 2023.

20. 'Sample Items from YSQ-L2 Long Form', Schema Therapy Institute, (n.d.); http://www.schematherapy.com/id53.htm, accessed 21 March 2023. Sample items from an earlier version of the Young Schema Questionnaire can be accessed on this site.

21. 'Sample Items from the YPI', Schema Therapy Institute, (n.d.); https://www.schematherapy.com/id205.htm (n.d.), accessed 21 March 2023. This is a subset of items from an earlier version of the Young Parenting Inventory.

22. Malwarebytes, 'Big Mother Is Watching'.

23. Michael Wesch, 'YouTube and You: Experiences of Self-awareness in the Context Collapse of the Recording Webcam', Explorations in Media Ecology, vol. 8, no. 2 (2009), pp. 19–34; http://hdl.handle.net/2097/6302, accessed 21 March 2023.

24. Barron Rodriguez et al., 'Remote Learning During the Global School Lockdown: Multi-Country Lessons', The World Bank, 30 March 2022; https://documents.worldbank.org/en/publication/documents-reports/documentdetail/668741627975171644/remote-learning-during-the-global-school-lockdown-multi-country-lessons, accessed 21 March 2023.

25. During Covid-19, digital divides deepened existing inequalities in society. Children with resources – connected devices, private rooms in which to study, technologically savvy parents, reliable and high-speed broadband, and access to the internet – were often able to continue and sometimes even enhance their education. The learning of children without these resources suffered greatly. See Victoria Coleman, 'Digital Divide in UK Education During COVID-19 Pandemic: Literature Review', Cambridge Assessment, 11 June 2021; https://www.cambridgeassessment.org.uk/Images/628843-digital-divide-in-uk-education-during-covid-19-pandemic-literature-review.pdf, accessed 21 March 2023.

26. Benjamin Herold, 'How Tech-Driven Teaching Strategies Have Changed During the Pandemic', Education Week, 14 April 2022; https://www.edweek.org/technology/how-tech-driven-teaching-strategies-have-changed-during-the-pandemic/2022/04, accessed 21 March 2023.

27. Emmeline Taylor, 'I Spy With My Little Eye: The Use of CCTV in Schools and the Impact on Privacy', The Sociological Review, vol. 58, no. 3 (20 July 2010), pp. 381–405; https://doi.org/10.1111/j.1467-954X.2010.01930.x, accessed 21 March 2023.

28. 'CCTV Model Policy for Schools', National Education Union (n.d); https://neu.org.uk/media/17481/view, accessed 21 March 2023.

29. Fraser Sampson, 'The State of Biometrics 2022: A Review of Policy and Practice in UK Education' (foreword), DefendDigitalMe, May 2022; https://defenddigitalme.org/research/state-biometrics-2022/, accessed 21 March 2023.

30. Crystal Tai, 'How China is Using Artificial Intelligence in Classrooms' (video), The Wall Street Journal, 1 October 2019; https://www.youtube.com/watch?v=JMLsHI8aVog, accessed 21 March 2023.

31. 'China Ranks "Good" and "Bad" Citizens with "Social Credit" System', France 24 English, 1 May 2019; https://www.youtube.com/watch?v=NXyzpMDtpSE, accessed 21 March 2023.

32. Samantha Hoffman's profile for the Australian Strategic Policy Initiative; https://www.aspi.org.au/bio/samantha-hoffman, accessed 21 March 2023.

33. Nicole Kobie, 'The Complicated Truth About China's Social Credit System', Wired, 6 July 2019; https://www.wired.co.uk/article/china-social-credit-system-explained, accessed 21 March 2023.

34. For ClassDojo's own statistics about usage and reach, see https://www.classdojo.com/en-gb/about/?redirect=true, accessed 21 March 2023.

35. L. Hooper, S. Livingstone and K. Pothong, 'Problems with Data Governance in UK Schools: The Cases of Google Classroom and ClassDojo', Digital Futures Commission/5Rights Foundation, 2022; https://digitalfuturescommission.org.uk/wp-content/uploads/2022/08/Problems-with-data-governance-in-UK-schools.pdf, accessed 21 March 2023.

36. ClassDojo, 'Privacy Policy', (n.d.); https://www.classdojo.com/en-gb/privacy/?redirect=true, accessed 21 March 2023. ClassDojo's privacy policy lists the following as being held by the app: Student data: '"Student Data" means any personal information, whether gathered by ClassDojo or provided by a school or its users, students, or students' parents/guardians for a school purpose, that is descriptive of the student including, but not limited to, information in the student's educational record or email, first and last name, birthdate, home or physical address, telephone number, email address, or other information allowing physical or online contact, discipline records, videos, test results, special education data, juvenile dependency records, grades, evaluations, criminal records, medical records, health records, social security numbers, biometric information, disabilities, socioeconomic information, food purchases, political affiliations, religious information text messages, documents, student identifies, search activity, photos, voice recordings, geolocation information, or any other information or identification number that would provide information about a specific student.'

37. 'Public Service Announcement: Education Technologies: Data Collection and Unsecured Systems Could Pose Risk to Students', Federal Bureau of Investigation, 13 September 2018; https://www.ic3.gov/Media/Y2018/PSA180913, accessed 21 March 2023. See also Kristen Walker et al., 'Compulsory Technology Adoption and Adaptation in Education: A Looming Student Privacy Problem', Journal of Consumer Affairs (28 December 2022); https://doi.org/10.1111/joca.12506, accessed 21 March 2023.

38. Jonathan Holmes, 'Schools Hit by Cyber Attack and Documents Leaked', BBC, 6 January 2023; https://www.bbc.co.uk/news/uk-england-gloucestershire-63637883, accessed 21 March 2023.

39. 'Privacy Evaluation for Google Classroom', Common Sense Privacy Program, 20 December 2022; https://privacy.commonsense.org/evaluation/Google-Classroom, accessed 21 March 2023.

40. Hooper et al., 'Problems with Data Governance in UK Schools'.

41. Personal communication, October 2022.

42. Siobhan Buchanan, 'An Apprehensive Teacher's Guide to . . . ClassDojo', Guardian, 28 November 2013; https://www.theguardian.com/teacher-network/teacher-blog/2013/nov/28/teacher-guide-classdojo-behaviour-management, accessed 21 March 2023.

43. Jachie Pohl, 'How ClassDojo Helped Me Become Teacher of the Year', The ClassDojo Blog, 12 July 2016; https://blog.classdojo.com/how-classdojo-helped-me-become-teacher-of-the-year, accessed 21 March 2023.

44. 'The Basics: Parent Accounts', ClassDojo information leaflet; https://static.classdojo.com/docs/TeacherResources/ParentFAQs/ParentFAQs_ForParents.pdf, accessed 21 March 2023.

45. A. Soroko, 'No Child Left Alone: The ClassDojo App', Our Schools/Our Selves, Spring 2018, pp. 63–75 (p. 70); https://www.researchgate.net/profile/Agata-Soroko/publication/304627527_No_child_left_alone_The_ClassDojo_app/links/5775

607508ae1b18a7dfdeed/No-child-left-alone-The-ClassDojo-app.pdf, accessed 21 March 2023.

46. Julie C. Garlen, 'ClassDojo raises concerns about children's rights' [online article], The Conversation, 25 February 2019; https://theconversation.com/classdojo-raises-concerns-about-childrens-rights-111033, accessed 30 June 2023.

47. Berni Graham, Clarissa White, Amy Edwards, Sylvia Potter and Cathy Street, 'School Exclusion: A Literature Review on the Continued Disproportionate Exclusion of Certain Children', Department for Education, May 2019; https://assets. publishing.service.gov.uk/government/ uploads/system/uploads/attachment_data/ file/800028/Timpson_review_of_school_ exclusion_literature_review.pdf, accessed 21 March 2023.

48. Hooper et al., 'Problems with Data Governance in UK Schools'.

49. See Safiya Noble, *Algorithms of Oppression: How Search Engines Reinforce Racism* (New York, NYU Press, 2018).

50. Personal communication, 25 January 2021.

51. Keith Fraser, 'Annual Statistics: A Youth Justice System Failing Black Children', Youth Justice Board for England and Wales, 27 January 2022; https://www.gov.uk/ government/news/annual-statistics-a-system-failing-black-children, accessed 21 March 2023.

52. 'Youth Justice Statistics: 2020 to 2021', Youth Justice Board of England and Wales, 27 January 2022; https:// www.gov.uk/government/statistics/ youth-justice-statistics-2020-to-2021/youth-justice-statistics-2020-to-2021-accessible-version, accessed 21 March 2023.

53. Drew Harwell, 'Federal Study Confirms Racial Bias of Many Facial-Recognition Systems, Casts Doubt on Their Expanded Use', *Washington Post*, 19 December 2019; https://www.washington post.com/technology/2019/12/19/ federal-study-confirms-racial-bias-many-facial-recognition-systems-casts-doubt-their-expanding-use/, accessed 21 March 2023.

This federal study in the US confirmed Facial Recognition Technology's propensity for racial bias, finding that Asian and African people were up to a hundred times more likely to be misidentified than white men, and that women were more likely to be falsely identified too. It's not the only study of its kind. Nevertheless, unquestioning belief in a fancy technology is hard to shake, and the promise of clearing a backlog of cold cases with minimal effort is like catnip to police forces.

54. Kashmir Hill, 'Wrongfully Accused by an Algorithm', *New York Times*, 3 August 2020; https://www.nytimes.com/2020/ 06/24/technology/facial-recognition-arrest. html, accessed 21 March 2023.

55. Ibid.

56. Tiffany L., Brown et al., 'Inclusion and Diversity Committee Report: What's Your Social Location?', National Council on Family Relations, 4 April 2019; https://www. ncfr.org/ncfr-report/spring-2019/inclusion-and-diversity-social-location, accessed 21 March 2023.

57. Drew Harwell, 'As Summer Camps Turn on Facial Recognition, Parents Demand: More Smiles, Please', *Washington Post*, 8 August 2019; https://www.washington post.com/technology/2019/08/08/ summer-camps-turn-facial-recognition-parents-demand-more-smiles-please/, accessed 21 March 2023.

Chapter 5: Adolescence

1. Amanda Silberling, 'TikTok Reached 1 Billion Monthly Active Users', Techcrunch. com, 27 September 2021; https://techcrunch. com/2021/09/27/tiktok-reached-1-billion-monthly-active-users, accessed 26 February 2023.

2. This statistic is provided by Netflix and was reported by the BBC. See Becky Padington and Ian Youngs, 'Anime: How Japanese Anime has Taken the West by Storm', BBC, 26 March 2022; https:// www.bbc.co.uk/news/entertainment-arts-60865649, accessed 26 February 2023.

3. Ibid.

4. Personal communication. Permission has been given for use of this interview material.

5. Tomura Shigaraki's *My Hero Academia* backstory is summarised in various places. See Francesco Cacciatore, '*My Hero Academia*'s Real Ultimate Villain Finally Takes the Stage', Screenrant.com, 30 January 2023, accessed 26 February 2023.

6. '*Sturm und Drang*: German Literary Movement', Britannica.com, 20 July 1998; https://www.britannica.com/event/Sturm-und-Drang/additional-info#history, accessed 26 February 2023.

7. Johann Wolfgang von Goethe, *The Sorrows of Young Werther* [1774], trans. R. D. Boylan, 2009; https://www.gutenberg.org/files/2527/2527-h/2527-h.htm, accessed 26 February 2023.

8. Andy Greenwald, *Nothing Feels Good: Punk Rock, Teenagers, and Emo* (New York, St Martin's Griffin, 2003).

9. Adam Burgess, 'A Guide to Goethe's "The Sorrows of Young Werther"', Thoughtco.com, 27 November 2018; https://www.thoughtco.com/sorrows-of-young-werther-goethe-739876, accessed 26 February 2023.

10. Charlie Connelly, 'This Book Conjured Seriously Sinister Reactions From Obsessed Fans', *The New European*, 13 March 2017; https://www.theneweuropean.co.uk/brexit-news-this-book-conjured-seriously-sinister-reactions-from-obsessed-fans-17088/, accessed 26 February 2023.

11. There is debate over whether the copycat suicides in Goethe's time were as numerous as rumoured. See Nicholas Kardaras, 'How Goethe's *Sorrows of Young Werther* Led to a Rare Suicide Cluster', Literary Hub, 15 September 2022; https://lithub.com/how-goethes-sorrows-of-young-werther-led-to-a-rare-suicide-cluster/, accessed 26 February 2023.

12. For an excellent discussion of the modern-day Werther Effect as weighed against the so-called Papageno effect – the idea that media and social media can provide support that bring potentially suicidal people back from the brink – see Jennifer Ouellette, 'How *13 Reasons Why* Sparked Years of Suicide-Contagion Research', *Ars Technica*, 15 September 2021; https://arstechnica.com/science/2021/09/how-13-reasons-why-sparked-years-of-suicide-contagion-research/, accessed 26 February 2023.

13. Granville Stanley Hall, *Adolescence: Its Psychology and its Relation to Physiology, Anthropology, Sociology, Sex, Crime, Religion and Education* (New York, D. Appleton and Company, 1904); https://archive.org/details/adolescenceitsp01hallgoog/page/n8/mode/2up, accessed 26 February 2023.

14. See Perry Meisel, 'Freudian Trip', *New York Times*, 24 January 1993; available on https://www.nytimes.com/1993/01/24/books/freudian-trip.html, accessed 26 February 2023. Perhaps Hall's move of bringing Freud and Jung to America was a career-enhancement strategy, for which he had to set his flagrantly anti-Semitic and distinctly anti-psychoanalytic views aside. Hall was a prominent eugenicist and his theories on the differences between the races and sexes might make his net legacy harmful rather than helpful.

15. Laurence A. Steinberg, 'A Social Neuroscience Perspective on Adolescent Risk-Taking', *Developmental Review*, vol. 28, no. 1 (March 2008), pp. 78–106; doi.org/10.1016/j.dr.2007.08.002. PMID: 18509515; PMCID: PMC2396566; https://www.ncbi.nlm.nih.gov/pmc/articles/PMC2396566/, accessed 26 February 2023.

16. Erikson and Erikson, *The Life Cycle Completed*. For a simple explanation of the Identity vs Role Confusion stage, see Mcleod, 'Erik Erikson's 8 Stages of Psychosocial Development'.

17. D. Wood et al., 'Emerging Adulthood as a Critical Stage in the Life Course' in N. Halfon, C. Forrest, R. Lerner and E. Faustman (eds), *Handbook of Life Course Health Development* (Springer, Cham, Switzerland, 2018); https://doi.org/10.1007/978-3-319-47143-3_7.

18. William Shakespeare, *As You Like It* (New York, Penguin Books, 2000).

19. For extensive coverage of the presentation of self online, with reference to Goffman and numerous other theorists, see Amy Guy, 'The Presentation of Self on a Decentralised Web', University of Edinburgh, doctoral thesis, 2018; https://dr.amy.gy, accessed on 26 February 2023.

20. Erikson and Erikson, *The Life Cycle Completed*.

21. This phrase is often attributed to Charles Darwin, but Darwin borrowed it from the sociologist and philosopher Herbert Spencer, who after reading Darwin's work used the phrase in his own book, *Principles of Biology* (1864). For more information, see Conor Cunningham, 'Survival of the Fittest – Biology'; https://www.britannica.com/science/survival-of-the-fittest (updated 16 February 2023), accessed 26 February 2023.

22. For a discussion of teens specifically within networked publics, see danah michele boyd, 'Taken Out of Context: Teen Sociality in Networked Publics', University of California Berkeley, doctoral thesis, 2008; https://www.danah.org/papers/TakenOutOfContext.pdf, accessed 26 February 2023.

23. Monica Anderson and Jingjing Jiang, 'Teens' Social Media Habits and Experiences', *Pew Research Center*, 28 November 2018; https://www.pewresearch.org/internet/2018/11/28/teens-social-media-habits-and-experiences/, accessed 26 February 2023.

24. Melanie C. Green and Jenna L. Clark, 'The Social Consequences of Online Interaction', in Alison Attrill-Smith, Chris Fullwood, Melanie Keep and Daria J. Kuss (eds), *The Oxford Handbook of Cyberpsychology* (Oxford, Oxford University Press, 2019), pp. 216–37 (p. 228).

25. J. P. Gerber, L. Wheeler and J. Suls, 'A Social Comparison Theory Meta-Analysis 60+ Years On', *Psychological Bulletin*, vol. 144, no. 2 (2018), pp. 177–97; https://doi.org/10.1037/bul0000127, accessed 18 March 2023.

26. S. Lyubomirsky and L. Ross, 'Hedeonic Consequences of Social Comparison: A Contrast of Happy and Unhappy People', *Journal of Personality and Social Psychology*, vol. 73, no. 6 (1997), pp. 1141–57; doi.org/10.1037//0022-3514.73.6.1141, accessed 19 March 2023.

27. Chia-chen Yang, Sean M. Holden, Mollie D. K. Carter and Jessica J. Webb, 'Social Media Social Comparison and Identity Distress at the College Transition: A Dual-Path Model', *Journal of Adolescence*, no. 69 (December 2018), pp. 92–102; https://doi.org/10.1016/j.adolescence.2018.09.007, accessed 19 March 2023.

28. 'The Panopticon', University College London (n.d.); https://www.ucl.ac.uk/bentham-project/who-was-jeremy-bentham/panopticon, accessed 26 February 2023.

29. Personal communication, used with permission.

30. Ben Bradford, Julia A. Yesberg, Jonathan Jackson and Paul Dawson, 'Live Facial Recognition: Trust and Legitimacy as Predictors of Public Support for Police Use of New Technology', *The British Journal of Criminology*, vol. 60, issue 6 (November 2020), pp. 1502–22; https://doi.org/10.1093/bjc/azaa032.

31. R. De Wolf, 'Contextualizing how teens manage personal and interpersonal privacy on social media', *New Media & Society*, vol. 22, no. 6 (2020), pp. 1058–75; https://doi.org/10.1177/1461444819876570.

32. 'The Facebook Files', *The Wall Street Journal*, September–December 2021; https://www.wsj.com/articles/the-facebook-files-11631713039, accessed 26 February 2023.

33. Georgia Wells, Jeff Horwitz and Deepa Seetharaman, 'Facebook Knows Instagram Is Toxic for Teen Girls, Company Documents Show', *The Wall Street Journal*, 14 September 2021; https://www.wsj.com/articles/facebook-knows-instagram-is-toxic-for-teen-girls-company-documents-show-11631620739?mod=article_inline, accessed 26 February 2023.

34. After the *Wall Street Journal*'s 'Facebook Files', Facebook released the annotated slides of the study results: https://about.fb.com/wp-content/uploads/2021/09/Instagram-Teen-Annotated-Research-Deck-1.pdf, accessed 26 February 2023.

35. Scott Pelley, 'Whistleblower: Facebook Is Misleading the Public on Progress Against Hate Speech, Violence, Misinformation', CBS News, 4 October 2021; https://www.cbsnews.com/news/facebook-whistleblower-frances-haugen-misinformation-public-60-minutes-2021-10-03/, accessed 26 February 2023.

36. Scott Carpenter, 'Zuckerberg Loses $6 Billion in Hours as Facebook Plunges', Bloomberg UK, 4 October 2021; https://www.bloomberg.com/news/articles/2021-10-04/zuckerberg-loses-7-billion-in-hours-as-facebook-plunges?leadSource=uverify%20wall, accessed 26 February 2023.

37. Avi Asher-Schapiro and Fabio Teixeira, 'Facebook Down: What the Outage Meant for the Developing World', Thomson Reuters Foundation News, 5 October 2021; https://news.trust.org/item/20211005204816-qzjft/, accessed 26 February 2023.

38. Carpenter, 'Zuckerberg Loses $6 billion'.

39. 'Facebook Whistleblower Frances Haugen Testifies Before Senate Commerce Committee', C-SPAN, 5 October 2021; https://www.youtube.com/watch?v=GOnpVQnv5Cw, accessed 26 February 2023.

40. Wells et al., 'Facebook Knows Instagram Is Toxic for Teen Girls'.

41. https://www.amyorben.com, accessed 26 February 2023.

42. Amy Orben, 'Teens, Screens and Well-Being: An Improved Approach', Oxford University, doctoral thesis, 2019; https://tinyurl.com/4tp8tadj, accessed 26 February 2023.

43. Ibid., p. 7.

44. For a layperson's explanation of the problems with self-report survey studies, see Kristalyn Salters-Pedneault, 'The Use of Self-Report Data in Psychology', Very Well Mind, 22 October 2022; https://www.verywellmind.com/definition-of-self-report-425267, accessed 26 February 2023.

45. M. J. George, M. R. Jensen, M. A. Russell, A. Gassman-Pines, W. E. Copeland, R. H. Hoyle and C. L. Odgers, 'Young Adolescents' Digital Technology Use, Perceived Impairments, and Well-Being in a Representative Sample', *Journal of Pediatrics*, vol. 219 (April 2020), pp. 180–7; doi.org/10.1016/j.jpeds.2019.12.002. Epub 2020 Feb 11. PMID: 32057438; PMCID: PMC7570431.

46. Leah Shafer, 'Social Media and Teen Anxiety', Harvard Graduate School of Education, 15 December 2017; https://www.gse.harvard.edu/news/uk/17/12/social-media-and-teen-anxiety, accessed 3 March 2023.

47. https://twitter.com/OrbenAmy.

48. Stuart Ritchie, 'Is Instagram Really Bad for Teenagers?', Unherd.com, 22 September 2021; https://unherd.com/2021/09/facebooks-bad-science/, accessed 26 February 2023.

49. Pratiti Raychoudhury, 'What Our Research Really Says about Teen Well-Being and Instagram' [online article], Facebook Newsroom, 26 September 2021; https://about.fb.com/news/2021/09/research-teen-well-being-and-instagram/, accessed 26 February 2023.

50. Natasha Lomas, 'Seeking to Respin Instagram's Toxicity for Teens, Facebook Publishes Annotated Slide Decks', TechCrunch.com, 30 September 2021; https://techcrunch.com/2021/09/30/seeking-to-respin-instagrams-toxicity-for-teens-facebook-publishes-annotated-slide-decks/, accessed 26 February 2023.

51. Casey Newton, 'Why Facebook Should Release the Facebook Files', TheVerge.com, 28 September 2021; https://www.theverge.com/22697786/why-facebook-should-release-the-facebook-files, accessed 26 February 2023.

52. Mariska Kleemans, Serena Daalmans, Ilana Carbaat and Doeschka Anschütz 'Picture Perfect: The Direct Effect of Manipulated Instagram Photos on Body Image in Adolescent Girls', *Media Psychology*, vol. 21, no. 1 (2018), pp. 93–110; doi.org/10.1080/15213269.2016.1257392.

53. 'Mental Health of Adolescents', World Health Organization, 17 November 2021; https://www.who.int/news-room/fact-sheets/detail/adolescent-mental-health, accessed 26 February 2023.

54. Lauren Feiner, 'Senate Will Grill Tech Execs After Report that Instagram Can Harm Teens' Mental Health', CNBC.com, 17 September 2021; https://www.cnbc.com/2021/09/17/senate-to-grill-tech-execs-after-report-on-instagram-teen-mental-health.html, accessed 26 February 2023.

55. For a simple explanation of the fragmentation and self-concept unity hypothesis of social media, see 'Who Am I, Facebook?', Association for Psychological Science, 19 April 2017; https://www.psychologicalscience.org/publications/observer/obsonline/who-am-i-facebook.html, accessed 27 February 2023.

56. Nick Yee and Jeremy Bailenson, 'The Proteus Effect: The Effect of Transformed Self-Representation on Behavior', *Human Communication Research*, vol. 33 (2007), pp. 271–90; doi.org/10.1111/j.1468-2958.2007.00299.x; https://stanfordvr.com/mm/2007/yee-proteus-effect.pdf, accessed 27 February 2023.

57. 'Who Am I, Facebook?', Association for Psychological Science.

58. Nandita Vijayakumar and Jennifer Pfeifer, 'Self-disclosure During Adolescence: Exploring the Means, Targets, and Types of Personal Exchanges', *Current Opinion in Psychology*, vol. 31 (February 2020), pp. 135–40; doi.org/10.1016/j.copsyc.2019.08.005, https://www.ncbi.nlm.nih.gov/pmc/articles/PMC7130455/, accessed 27 February 2023.

59. Emily Towner, Jennifer Grint, Tally Levy, Sarah-Jayne Blakemore and Livia Tomova, 'Revealing the Self in a Digital World: A Systematic Review of Adolescent Online and Offline Self-Disclosure', *Current Opinion in Psychology* (June 2022), 45:101309; doi.org/10.1016/j.copsyc.2022.101309; epub, 18 February 2022; https://pubmed.ncbi.nlm.nih.gov/35325809/, accessed 27 February 2023.

60. Amy Orben, 'Don't Despair if Your Child is Glued to a Screen, It May be Keeping them Sane', *Guardian*, 26 April 2020; https://www.theguardian.com/commentisfree/2020/apr/26/dont-despair-if-your-child-is-glued-to-a-screen-it-may-be-keeping-them-sane, accessed 27 February 2023.

61. 'There is nothing either good or bad, but thinking makes it so', *Hamlet*, Act II scene 2, spoken by Hamlet.

62. Federica Pedalino and Anne-Linda Camerini, 'Instagram Use and Body Dissatisfaction: The Mediating Role of Upward Social Comparison with Peers and Influencers among Young Females', *International Journal of Environmental Research and Public Health*, vol. 19, issue 3 (29 January 2022), article 1543; https://doi.org/10.3390/ijerph19031543, accessed 3 March 2023.

63. Green and Clark, 'The Social Consequences of Online Interaction'.

64. Jaimee Stuart, Riley Scott, Karlee O'Donnell and Paul E. Jose, 'The Impact of Ease of Online Self-Expression During Adolescence on Identity in Young Adulthood', in T. Machin, C. Brownlow, S. Abel and J. Gilmour (eds), *Social Media and Technology Across the Lifespan: Palgrave Studies in Cyberpsychology* (London, Palgrave Macmillan, 2022); https://doi.org/10.1007/978-3-030-99049-7_4.

65. Erikson and Erikson, *The Life Cycle Completed*.

66. Erving Goffman, *The Presentation of Self in Everyday Life* (New York, NY, Bantam Doubleday Dell, 1959).

67. Paul Mountfort, Anne Peirson-Smith and Adam Geczy, *Planet Cosplay: Costume Play, Identity and Global Fandom* (Chicago, University of Chicago Press, 2018).

68. M. H. Bornstein and D. Zlotnik, 'Parenting Styles and Their Effects', in Marshall M. Haith and Janette B. Benson (eds), *Encyclopedia of Infant and Early Childhood Development*, 2nd edn (Cambridge, MA, Elsevier Academic Press, 2008), pp. 496–509.

69. Orben, https://tinyurl.com/4tp8tadj.

70. Mountfort et al., *Planet Cosplay*.

71. Luciano Floridi (ed.), *The Online Manifesto: Being Human in a Hyperconnected Era* (London, Springer Open, 2015). The eBook is available for download from https://link.springer.com/book/

10.1007/978-3-319-04093-6, accessed 26 February 2023.

72. Ibid., p. 1.

73. William Gibson, *Neuromancer* (London, Harper Voyager, 1984).

74. Patrick Stokes, *Digital Souls: A Philosophy of Online Death* (London, Bloomsbury Academic Publishing, 2021).

Chapter 6: Adulthood

1. Erikson and Erikson, *The Life Cycle Completed*. For a simple explanation of the Intimacy vs Isolation stage, see Mcleod, 'Erik Erikson's 8 Stages of Psychosocial Development'.

2. Alex Shashkevich, 'Meeting online has become the most popular way U.S. couples connect, Stanford sociologist finds', *Stanford News*, 21 August 2019; https://news.stanford.edu/2019/08/21/online-dating-popular-way-u-s-couples-meet/, accessed 4 March 2023.

3. Samantha Joel et al., 'Machine learning uncovers the most robust self-report predictors of relationship quality across 43 longitudinal couples studies', *PNAS/ Psychological and Cognitive Sciences*, vol. 117, no. 32 (11 August 2020), 19061–71; https://doi.org/10.1073/pnas.1917036117 (27 July 2020), accessed 4 March 2023.

4. Alex Cocotas, 'History Tells Us – 685 Million Smartphones Will Be Sold This Year – A 45 Percent Increase Over Last Year', BusinessInsider.com, 14 August 2012; https://www.businessinsider.com/what-history-tells-us-about-2012-smartphone-sales-2012-8?r=US&IR=T, accessed 4 March 2023.

5. Tom Chatfield, 'The new words that expose our smartphone obsessions', BBC.com, 29 November 2016; https://www.bbc.com/future/article/20161129-the-new-words-that-reveal-how-tech-has-changed-us, accessed 4 March 2023.

6. Elyssa M. Barrick, Alixandria Barasch and Diana I. Tamir, 'The Unexpected Social Consequences of Diverting Attention to Our Phones', *Journal of Experimental Social Psychology*, vol. 101 (July 2022); https://doi.org/10.1016/j.jesp.2022.104344, accessed 4 March 2023.

7. Yeslam Al-Saggaf and Sarah B. O'Donnell, 'Phubbing: Perceptions, Reasons Behind, Predictors, and Impacts', Wiley Online Library: *Human Behavior and Emerging Technologies*, 10 April 2019; https://doi.org/10.1002/hbe2.137, accessed 4 March 2023.

8. Camiel J. Beukeboom and Monique Pollmann, 'Partner Phubbing: Why Using Your Phone During Interactions with Your Partner Can Be Detrimental for Your Relationship', *Computers in Human Behavior*, vol. 124 (November 2021); https://doi.org/10.1016/j.chb.2021.106932, accessed 4 March 2023.

9. Tarja Salmela, Ashley Colley and Jonna Häkkilä, 'Together in Bed? Couples' Mobile Technology Use in Bed', *Proceedings of the CHI Conference on Human Factors in Computing Systems*, paper no. 502 (May 2019), pp. 1–12; https://doi.org/10.1145/3290605.3300732, accessed 4 March 2023.

10. See, for example, Shalini Misra, Lulu Cheng, Jamie Genevie and Miao Yuan, 'The iPhone Effect: The Quality of In-Person Social Interactions in the Presence of Mobile Devices', *Environment and Behavior*, vol. 48, issue 2 (1 July 2014); https://journals.sagepub.com/doi/10.1177/0013916514539755; accessed 4 March 2023.

11. Mariek M. P. Vanden Abeele and Marie Postma-Nilsenova, 'More Than Just Gaze: An Experimental Vignette Study Examining How Phone-Gazing and Newspaper-Gazing and Phubbing-While-Speaking and Phubbing-While-Listening Compare in Their Effect on Affiliation', *Communication Research Reports*, vol. 35, issue 4 (2018), pp. 303–13; https://doi.org/10.1080/08824096.2018.1492911, 17 July 2018, accessed 4 March 2023.

12. For more information on these concepts as used within communications privacy management (CPM) theory, see the Communication Privacy Management Center website on https://cpmcenter.iupui.edu/learn/theory, accessed 4 March 2023.

13. Charlotte D. Vinkers, C. Finkenauer and Skyler T. Hawk, 'Why Do Close Partners Snoop? Predictors of Intrusive Behavior in Newlywed Couples', *Personal Relationships*, vol. 18, issue 1 (March 2011), pp. 110–24; https://doi.org/10.1111/j.1475-6811.2010.01314.x, 11 November 2010, accessed 4 March 2023.

14. See Amir Levine and Rachel Heller, *Attached: Are You Anxious, Avoidant or Secure? How the Science of Adult Attachment Can Help You Find – and Keep – Love* (New York, Pan Macmillan/Bluebird, 2019). For a robust researched discussion of attachment theory and styles, visit R. Chris Fraley, 'Adult Attachment Theory and Research: A Brief Overview'; http://labs.psychology.illinois.edu/~rcfraley/attachment.htm (2018), accessed 4 March 2023.

15. Vinkers et al., 'Why do close partners snoop?'

16. Ibid.

17. Jesse Fox and Robert S. Tokunaga, 'Romantic Partner Monitoring After Breakups: Dependence, Distress, and Post-Dissolution Surveillance via Social Networking Sites', *Cyberpsychology, Behavior, and Social Networking*, vol. 18, no. 9 (8 September 2015); doi.org/10.1089/cyber.2015.0123 (8 September 2015), accessed 4 March 2023.

18. 'What is Stalking and Harassment?', Police UK (n.d); https://www.police.uk/advice/advice-and-information/sh/stalking-harassment/what-is-stalking-harassment/, accessed 4 March 2023.

19. 'Stalking and Harassment', Crown Prosecution Service, 23 May 2018; https://www.cps.gov.uk/legal-guidance/stalking-and-harassment, accessed 4 March 2023.

20. '"Fighting for My Sanity": Stalking and Post-Traumatic Stress Disorder', Suzy Lamplugh Trust, April 2019; https://www.suzylamplugh.org/fighting-for-my-sanity-stalking-and-post-traumatic-stress-disorder, accessed 4 March 2023.

21. Ibid.

22. 'Stalking Analysis Reveals Domestic Abuse Link', Crown Prosecution Service, 4 December 2020; https://www.cps.gov.uk/cps/news/stalking-analysis-reveals-domestic-abuse-link, accessed 4 March 2023.

23. The definition of coercive control includes a personal connection between victim and perpetrator; the victim fearing violence will be used against them or experiencing a significant adverse effect on their lives; behaviour that is repeated or continuous; and the perpetrator's likely knowledge that their behaviour would have a serious effect on the victim. See https://www.gov.uk/government/consultations/controlling-or-coercive-behaviour-statutory-guidance/draft-controlling-or-coercive-behaviour-statutory-guidance-accessible, 23 April 2022, accessed 4 March 2023.

24. Emma Short, Antony Brown, Melanie Pitchford and James Barnes, 'Revenge Porn: Findings from the Harassment and Revenge Porn (HARP) Survey – Preliminary Results', Annual Review of Cybertherapy and Telemedicine, 2017; https://pure.uvt.nl/ws/portalfiles/portal/24446962/Annual_Review_of_CyberTherapy_and_Telemedicine_2017.pdf#page=208, accessed 4 March 2023.

25. Samantha Bates, 'Revenge Porn and Mental Health: A Qualitative Analysis of the Mental Health Effects of Revenge Porn on Female Survivors', *Feminist Criminology*, vol. 12, issue 1 (24 July 2016); https://doi.org/10.1177/1557085116654565, accessed 4 March 2023.

26. '2021 Norton Cyber Safety Insights Report: Special Release – Online Creeping', Norton Lifelock; https://www.nortonlifelock.com/us/en/newsroom/press-kits/2021-norton-cyber-safety-insights-report-special-release-online-creeping/, accessed 1 November 2022. For summary, see '2021 Norton Cyber Safety Insights Report: Special Release – Online Creeping', YouTube, 14 June 2021; https://www.youtube.com/watch?v=H1gWXwDKOd8, accessed 4 March 2023.

27. Emily Tseng et al., 'The Tools and Tactics Used in Intimate Partner Surveillance: An Analysis of Online Infidelity Forums', 29th Usenix Security Symposium 12–14 August 2020; https://www.usenix.org/

conference/usenixsecurity20/presentation/tseng, accessed 4 March 2023.

28. 'Am I Being Stalked?', National Stalking Helpline (online assessment tool) (n.d.); https://www.suzylamplugh.org/am-i-being-stalked-tool, accessed 4 Mar 2023.

29. For raw data on stalking in England and Wales from 2022, see Office of National Statistics, 'Stalking: findings from the Crime Survey for England and Wales' (dataset); https://www.ons.gov.uk/peoplepopulation andcommunity/crimeandjustice/datasets/stalkingfindingsfromthecrimesurveyfor englandandwales 25 November 2022, accessed 4 March 2023.

30. Crown Prosecution Service, 'Stalking Analysis Reveals Domestic Abuse Link'.

31. 'Understanding and addressing violence against women: Intimate partner violence', World Health Organization, 2012; https://apps.who.int/iris/bitstream/handle/10665/77432/WHO_RHR_12.36_eng.pdf, accessed 4 March 2023.

32. 'About Us', Clinic to End Tech Abuse (CETA), 2023; https://www.ceta.tech.cornell.edu/aboutus, accessed 4 March 2023.

33. Crown Prosecution Service, 'Stalking and Harassment'.

34. 'Am I Being Stalked?', Stalking in Ireland (n.d.); https://www.stalkingin ireland.ie/am-i-being-stalked, accessed 4 March 2023.

35. Norton Lifelock, '2021 Norton Cyber Safety Insights Report'.

36. 'GenZ and Millennials Accept Online Creeping and Stalking as Part of Dating Culture', PRNewswire, 7 February 2023; https://www.prnewswire.com/news-releases/gen-z-and-millennials-accept-online-creeping-and-stalking-as-part-of-dating-culture-301740012, accessed 4 March 2023.

37. 'Written Evidence from the Suzy Lamplugh Trust' (evidence to parliamentary committee), UK Parliament, 5 June 2022; https://committees.parliament.uk/written evidence/109132/pdf/, accessed 4 March 2023.

38. 'BLOG: Stalking During Covid-19', Victims Commissioner, 28 July 2021; https://victimscommissioner.org.uk/news/blog-stalking-during-covid-19/, accessed 4 March 2023.

39. Catherine Lough, 'Six-Fold Increase to Incel Websites in Nine Months', *Independent*, 3 January 2022; https://www.independent.co.uk/news/uk/jake-davison-plymouth-law-commission-ofcom-b1985878.html, accessed 4 March 2023.

40. Emily Tseng et al., 'The Tools and Tactics Used in Intimate Partner Surveillance'.

Chapter 7: Middle Adulthood

1. Mary Lewis, 'Frame Knitting', *Heritage Crafts: Red List of Endangered Crafts*, 24 May 2021; https://heritagecrafts.org.uk/frame-knitting/, accessed 6 March 2023.

2. Jessica Brain, 'The Luddites', Historic UK: The History and Heritage Accommodation Guide, 6 October 2018; https://www.historic-uk.com/HistoryUK/HistoryofBritain/The-Luddites/, accessed 6 March 2023.

3. Tom De Castella, 'Are You a Luddite?', BBC.co.uk, 20 April 2012; https://www.bbc.co.uk/news/magazine-17770171, accessed 6 March 2023.

4. Thomas Pynchon, 'Is it O.K. to Be a Luddite?', *New York Times Book Review*, 28 October 1984, pp. 1, 40–1; http://www.pynchon.pomona.edu/uncollected/luddite.html, accessed 6 March 2023.

5. Marc Prensky, 'Digital Natives, Digital Immigrants, Part 1', *On the Horizon*, vol. 9, issue 5 (1 September 2001), pp. 1–6; http://dx.doi.org/10.1108/10748120110424816, accessed 7 March 2023.

6. Erikson and Erikson, *The Life Cycle Completed*. For a simple explanation of the Generativity vs Stagnation stage, see Mcleod, 'Erik Erikson's 8 Stages of Psychosocial Development'.

7. Harold Ramis (director), *Groundhog Day* (Columbia Pictures, 1993).

8. Rachel Nuwer, 'Our Personalities are Most Stable in Mid-Life', *Smithsonian Magazine*, 14 August 2014; https://www.smithsonianmag.com/smart-news/

our-personalities-are-most-stable-mid-life-180952352/, accessed 7 March 2023.

9. Colleen McClain, Emily A. Vogels, Andrew Perrin, Stella Sechopoulos and Lee Rainie, 'The Internet and the Pandemic', Pew Research Center, 1 September 2021; https://www.pewresearch.org/internet/2021/09/01/the-internet-and-the-pandemic/, accessed 7 March 2023.

10. 'Covid-19 and the Digital Divide', UK Parliament, 17 December 2020; https://post.parliament.uk/covid-19-and-the-digital-divide/, accessed 7 March 2023.

11. To take a psychometric assessment for the Big Five, visit the Open-Source Psychometrics Project on https://open psychometrics.org/tests/IPIP-BFFM, accessed 18 August 2022. For HEXACO, go to http://hexaco.org, accessed 7 March 2023.

12. Maureen Tibbetts, Adam Epstein-Shuman, Matthew Leitao and Kostadin Kushlev, 'A Week During COVID-19: Online Social Interactions are Associated with Greater Connection and More Stress', *Computers in Human Behavior*, vol. 4 (August–December 2021); https://doi.org/10.1016/j.chbr.2021.100133, accessed 7 March 2023.

13. Ave Kotze, 'The Therapeutic Frame: An Illusion of the Past?', *The Psychologist*, 28 February 2023; https://www.bps.org.uk/psychologist/therapeutic-frame-illusion-past, accessed 7 March 2023.

14. L. C. Jiang, N. N. Bazarova and J. T. Hancock (2013), 'From Perception to Behavior: Disclosure Reciprocity and the Intensification of Intimacy in Computer-mediated Communication', *Communication Research*, vol. 40, issue 1 (February 2013), pp. 125–43.

15. Elaine Kasket, 'The Digital Age Technologies Attitudes Scale (DATAS)'; https://tinyurl.com/58pnb753. Check https://www.elainekasket.com for updates and scoring.

16. David S. White and Alison Le Cornu, 'Visitors and Residents: A New Typology for Online Engagement', *First Monday*, vol. 16, no. 9 (5 September 2011); https://doi.org/10.5210/fm.v16i9.3171, accessed 7 March 2023.

17. J. L. Clark and M. C. Green, 'Self-fulfilling prophecies: Perceived reality of online interaction drives expected outcomes of online communication', *Personality and Individual Differences* (2017), advanced online publication; doi.org/10.1016/j.paid.2017.08.031.

18. David Scharff, 'The Data Are In: Telehealth is Here to Stay', *Psychology Today*, 10 November 2021; https://www.psychologytoday.com/gb/blog/psychoanalytic-exploration/202111/the-data-are-in-telehealth-is-here-stay, accessed 7 March 2023.

19. Elaine Kasket, 'When Virtual Becomes Reality', *IAI News*, issue 87 (9 April 2020); https://iai.tv/articles/when-virtual-becomes-reality-auid-1402, accessed 7 March 2023.

20. Debra Rose Wilson (reviewer) and Louise Morales-Brown (author), 'What Does It Mean to be "Touch Starved"?', *Medical News Today*, 19 January 2021; https://www.medical newstoday.com/articles/touch-starved, accessed 7 March 2023.

21. Rosetta Thurman, 'Why We Love Social Media: Could Findings from a Study on Oxytocin be Applied to Social Media?', Stanford Social Innovation Review, 20 July 2010; https://ssir.org/articles/entry/why_we_love_social_media, accessed 7 March 2023.

22. Paul Zak, 'How Stories Change the Brain', *Greater Good Magazine*, 17 December 2013; https://greatergood.berkeley.edu/article/item/how_stories_change_brain, accessed 7 March 2023.

23. NCFE, 'Digital poverty: 3 Factors and How Society Can Tackle It', NCFE.org.uk; https://www.ncfe.org.uk/all-articles/digital-poverty-3-factors-and-how-society-can-tackle-it/, accessed 7 March 2023.

24. Ibid.

25. 'Poorest Twice as Likely to Feel Lonely in Lockdown Compared to the Richest', University of Manchester, 28 July 2021; https://www.manchester.ac.uk/discover/news/poorest-twice-as-likely-to-feel-lonely-in-lockdown/, accessed 7 March 2023.

26. Edmund Heery and Mike Noon, *A Dictionary of Human Resource Management*,

2nd edn (Oxford, Oxford University Press, 2008); doi.org/10.1093/acref/9780199298 761.001.0001, accessed 6 March 2023. See summary at https://www.oxford reference.com/display/10.1093/oi/authority .20110803100021241;jsessionid=AA35ACDC B4B4A61C5A0A844E566C3FF2, accessed 6 March 2023.

27.   Alex De Ruyter and Martyn Brown, *The Gig Economy* (Newcastle-upon-Tyne, Agenda Publishing, 2019).

28.   'The Origins of "Side-Hustle": The Word Appears to Have Entered Our Language in the 1950s', Merriam-Webster/ Words We're Watching, September 2022; https://www.merriam-webster.com/words-at-play/words-were-watching-side-hustle, accessed 6 March 2023. It's only in the last couple of decades, apparently, that 'side hustle' has come to be used in the 'job' sense of the word – throughout much of its history, it had more unsavoury connotations, i.e., 'scam'.

29.   Nikki Shaner-Bradford, 'What Do you Do? I'm a Podcaster-Vlogger-Model-DJ: The Gig Economy Has Given Us Multi-Hyphenates: People Who've Made a Lifestyle Out of Working More than One Job', The Outline.com, 25 November 2019; https://theoutline.com/post/8301/everyone-you-know-is-a-multi-hyphenate, accessed 6 March 2023.

30.   FutureLearn, 'Global Report Suggest "Job for Life" a Thing of the Past', FutureLearn.com, 22 April 2021; https:// www.futurelearn.com/info/press-releases/ global-report-suggests-job-for-life-a-thing-of-the-past, accessed 6 March 2023.

31.   McKinsey Global Institute, 'The Future of Work After COVID-19', 18 February 2021; https://www.mckinsey. com/featured-insights/future-of-work/ the-future-of-work-after-covid-19, accessed 6 March 2023.

32.   Raj Raghunathan, *If You're So Smart, Why Aren't You Happy?* (New York, Penguin/ Portfolio, 2016).

33.   *Modern Times*, directed by Charlie Chaplin (United Artists, 1936).

34.   Raghunathan, *If You're So Smart*.

35.   David Blustein, 'The Great Resignation', Boston College, 10 April 2023; https:// www.bc.edu/bc-web/bcnews/nation-world-society/education/q-a-the-great-resignation. html, accessed 7 March 2023.

36.   Dan Cable and Lynda Gratton, 'What's Driving the Great Resignation?', Think/London Business School, 13 June 2022; https://www.london.edu/think/whats-driving-the-great-resignation, accessed 7 Mar 2023.

37.   Joseph Fuller and William Kerr, 'The Great Resignation Didn't Start with the Pandemic', *Harvard Business Report*, 23 March 2022; https://hbr.org/2022/03/ the-great-resignation-didnt-start-with-the-pandemic, accessed 7 March 2023.

38.   'What 52,000 People Think About Work Today: PwC's Global Workforce Hopes and Fears Survey 2022', PWC Global, 24 May 2022; https://www.pwc.com/gx/en/ issues/workforce/hopes-and-fears-2022. html, accessed 7 March 2022.

39.   Ibid.

40.   Nadia Reckmann, '4 Ways to Implement Peter Drucker's Theory of Management', *Business News Daily*, 21 February 2023; https://www.business newsdaily.com/10634-peter-drucker-management-theory.html, accessed 7 March 2023.

41.   Sébastien Ricard, 'The Year of the Knowledge Worker', Forbes, 10 December 2020; https://www.forbes.com/sites/forbes techcouncil/2020/12/10/the-year-of-the-knowledge-worker/?sh=597b73907fbb, accessed 7 March 2023.

42.   McKinsey & Company, 'Americans are embracing Flexible Work – and They Want More of It', McKinsey.com, 23 June 2022; https://www.mckinsey. com/industries/real-estate/our-insights/ americans-are-embracing-flexible-work-and-they-want-more-of-it, accessed 7 March 2023.

43.   Reckmann, '4 Ways to Implement Peter Drucker's Theory of Management'.

44.   Gabriella Swerling, 'Still Working from Home? You're Probably Middle-Aged and Wealthy', *Telegraph*, 23 May 2022; https:// www.telegraph.co.uk/news/2022/05/

23/still-working-home-probably-middle-aged-wealthy/, accessed 7 March 2023.

45. Robbert van Eerd and Jean Guo, 'Jobs Will Be Very Different in Ten Years. Here's How to Prepare', World Economic Forum: Future of Work, 17 January 2020; https://www.weforum.org/agenda/2020/01/future-of-work/, accessed 7 March 2023.

46. 'Ford to Cut Nearly 4,000 Jobs in Europe, Including 1,300 in UK: US Carmaker Blames Losses on Rising Costs and Need to Switch to Electric Vehicle Production', *Guardian*, 14 February 2023; https://www.theguardian.com/business/2023/feb/14/ford-jobs-losses-cut-uk-europe, accessed 7 March 2023.

47. 'Hospice Chaplain Demographics and Statistics in the US', Zippia; https://www.zippia.com/hospice-chaplain-jobs/demographics/, accessed 7 March 2023.

48. Jodi Kantor and Arya Sundaram, 'The Rise of the Worker Productivity Score', *New York Times*, 14 August 2022; https://www.nytimes.com/interactive/2022/08/14/business/worker-productivity-tracking.html, accessed 7 March 2023.

49. Ibid.

50. Ibid.

51. Ibid.

52. William Alexander, *Ten Tomatoes that Changed the World: A History* (New York, Little, Brown & Company, 2022).

53. Michael Barbaro (host), Rikki Novetsy, Michael Simon Johnson and Mooj Zadie (producers), 'The Rise of Workplace Surveillance', *The Daily* (podcast), 24 August 2022; https://www.nytimes.com/2022/08/24/podcasts/the-daily/workplace-surveillance-productivity-tracking.html, accessed 7 March 2023.

54. Rose, Nikolas, *Governing the Soul: Shaping of the Private Self* (London, Free Association Publishing, 1999).

55. https://houseofbeautifulbusiness.com, accessed 7 March 2023.

56. Rose, *Governing the Soul*.

57. Douglas Belkin and Lindsay Ellis, 'Mouse Jigglers, Fake PowerPoints: Workers Foil Bosses' Surveillance Attempts: Companies that Track Employees'

Productivity Run Up Against Their Inventive Workarounds', *The Wall Street Journal*, 11 January 2023; https://www.wsj.com/articles/employee-productivity-workers-avoid-monitoring-11673445411, accessed 7 March 2023.

58. Chase Thiel, Julena M. Bonner, John Bush, David Welsh and Niharika Garud, 'Monitoring Employees Makes Them More Likely to Break Rules', *Harvard Business Review*, 27 July 2022; https://hbr.org/2022/06/monitoring-employees-makes-them-more-likely-to-break-rules, accessed 7 March 2023.

59. Ibid.

60. Kantor and Sundaram, 'The Rise of the Worker Productivity Score'.

61. This Krishnamurti quote is apocryphal, assigned to him by Mark Vonnegut in *The Eden Express: A Memoir of Insanity* (1975). It probably has its origin in passages from Krishnamurti's *Commentaries on Living Series* (3rd series, 1960): 'Has not society itself helped to make the individual unhealthy? . . . why should the individual adjust himself to an unhealthy society?'

62. Personal communication.

63. Emily Laurence, 'Virtual Reality Therapy: Everything You Need to Know', *Forbes Health*, 31 January 2023; https://www.forbes.com/health/mind/virtual-reality-therapy/, accessed 7 March 2023.

64. Stephen L. Carter, 'Can ChatGPT Write a Novel Better Than I Can?', *Washington Post*, 13 February 2023; https://www.washingtonpost.com/business/can-chatgpt-write-a-better-novel-than-i-can/2023/02/11/e7f3d0c4-aa0e-11ed-b2a3-edb05ee0e313_story.html, accessed 7 March 2023.

65. Anonymous, personal communication, 24 March 2023.

66. Laurie Clarke, 'When AI Can Make Art – What Does It Mean for Creativity?', *Guardian*, 12 November 2022; https://www.theguardian.com/technology/2022/nov/12/when-ai-can-make-art-what-does-it-mean-for-creativity-dall-e-midjourney, accessed 7 March 2023.

67. University and College Union, '"Hands Off" Online Lecture Recordings,

UCU Tells Universities and Colleges', UCU. org.uk, 3 September 2021; https://www. ucu.org.uk/article/11746/Hands-off-onlin e-lecture-recordings-union-tells-universit ies-and-colleges, accessed 7 March 2023.

68. Tamara Kneese, 'How a Dead Professor is Teaching a University Art History Class', *Slate: FutureTense*, 27 January 2021; https://slate.com/technology/2021/01/ dead-professor-teaching-online-class.html, accessed 7 March 2023.

69. Stephen Marche, 'The College Essay Is Dead: Nobody Is Prepared for How AI Will Transform Academia', *The Atlantic*, 6 December 2022; https://www.theatlantic. com/technology/archive/2022/12/chatgpt- ai-writing-college-student-essays/672371/, accessed 7 March 2023.

Chapter 8: Older Adulthood

1. Statista Research Department, 'Average Cost of a 30-Second Super Bowl TV Commercial in the United States from 2002 to 2023', Statista.com, 10 February 2023; https://www.statista.com/statistics/217134/ total-advertisement-revenue-of-super-bowls/, accessed 3 March 2023.

2. Jordan Valinsky, 'Google's Super Bowl Ad: People Shed Tears for "Loretta"', CNN Business, 3 February 2020; https:// edition.cnn.com/2020/02/03/business/ google-super-bowl-ad-loretta-trnd/index. html, accessed 3 March 2023.

3. Justin Byers, '112.3 Million People Watched Super Bowl LVI', FrontOfficeSports. com, 15 February 2022; https://frontoffice sports.com/112-million-people-watched- super-bowl-lvi/, accessed 3 March 2023.

4. Kayla Keegan, 'The Incredible True Story Behind Google's 2020 Super Bowl Commercial: "Loretta" Was One of the Most Powerful Ads of the Night', GoodHousekeeping.com, 3 February 2020; https://www.goodhousekeeping.com/life/ entertainment/a30751020/super-bowl- google-commercial-ad-true-story/, accessed 3 March 2023.

5. Megan McCluskey and Rachel E. Greenspan, 'These Were the Best Super Bowl

2020 Commercials', Time.com, 2 February 2020; https://time.com/5772692/best- super-bowl-commercials-2020/, accessed 3 March 2023.

6. Lorraine Twohill, 'Google and the Super Bowl: Here to Help', Google: The Keyword, 28 January 2020; https://blog. google/products/assistant/google-super- bowl-here-to-help/, accessed 3 March 2023.

7. Joelle Renstrom, 'The Sinister Realities of Google's Tear-Jerking Super Bowl Commercial', Slate.com, 3 February 2020; https://slate.com/technology/2020/02/ google-assistant-super-bowl-commercial- loretta.html, accessed 3 March 2023.

8. Erikson and Erikson, *The Life Cycle Completed*. For a simple explanation of the Ego Integrity vs Despair stage, see Mcleod, 'Erik Erikson's 8 Stages of Psychosocial Development'.

9. For a layperson's explanation, see Saul Mcleod, 'Freud's Id, Ego, and Superego: Definition and Examples', Simply Psychology, 20 February 2023; https://simply psychology.org/psyche.html, accessed 3 March 2023.

10. Elaine Kasket, 'Memento Mori, Memento Vivere: Use an Exercise About Death to Help You Live a Better Life', Ascent Publication/Medium.com, 7 November 2019; https://medium.com/the-ascent/memento- mori-memento-viveri-8df26bd2aad8, accessed 3 March 2023.

11. 'Mom Computer Therapy' (comedy sketch), *Inside Amy Schumer* (television series), season 2, episode 7, Amy Schumer, Brooke Posch and Daniel Powell (executive producers), So Easy Productions/ Irony Point/Jax Media (2014); https://www. youtube.com/watch?v=A6A331B10q8, accessed 3 March 2023.

12. Kelly Twohig, 'Why Marketers' Picture of Seniors is Getting Old', ThinkWithGoogle: Consumer Insights, July 2021; https://www.thinkwithgoogle.com/ consumer-insights/consumer-trends/digital- seniors/, accessed 3 March 2023.

13. 'IT and Internet Industry', Office of National Statistics (data and analysis from Census 2021); https://www.ons.gov.

uk/businessindustryandtrade/itandinternet industry, accessed on 3 March 2023.

14. Data is still emerging, but see, for example, Andrew Sixsmith, Becky R. Horst, Dorina Simeonov and Alex Mihailidis, 'Older People's Use of Digital Technology During the COVID-19 Pandemic', *Bulletin of Science, Technology & Society*, vol. 42, issue 1–2 (21 April 2022); doi.org/10.1177/02704676221094731, accessed 3 March 2023.

15. Twohig, 'Why marketers' picture of seniors is getting old'.

16. 'Build a Helpful Home, One Device at a Time', Google Home, 2023; https://home.google.com/what-is-google-home/, accessed 3 March 2023.

17. Viktor Mayer-Schönberger, *Delete: The Virtue of Forgetting in the Digital Age* (Cambridge, MA, Princeton University Press, 2009).

18. 'Dementia', World Health Organization, 20 September 2022; https://www.who.int/news-room/fact-sheets/detail/dementia, accessed 3 March 2023.

19. Anne Wilson and Michael Ross, 'The Identity Function of Autobiographical Memory: Time is On Our Side', *Memory*, vol. 11, issue 2 (21 October 2010), pp. 137–49; doi.org/10.1080/741938210, accessed 3 March 2023.

20. Michael K. Scullin, Winston E. Jones, Richard Phenis, Samantha Beevers, Sabra Rosen, Kara Dinh, Andrew Kiselica, Francis J. Keefe and Jared F. Benge, 'Using Smartphone Technology to Improve Prospective Memory Functioning: A Randomized Controlled Trial', *Journal of the American Geriatrics Society*, vol. 70, issue 2 (February 2022), pp. 459–69; https://doi.org/10.1111/jgs.17551, accessed 3 March 2023.

21. Samantha A. Wilson, Paula Byrne, Sarah E. Rodgers and Michelle Maden, 'A Systematic Review of Smartphone and Tablet use by Older Adults With and Without Cognitive Impairment', *Innovation in Aging*, vol. 6, issue 2 (6 January 2022); doi.org/10.1093/geroni/igac002, accessed 3 March 2023.

22. Eidetic is a spaced-repetition learning app, which helps memorise phone numbers, words and facts; https://www.eideticapp.com, accessed 3 March 2023.

23. Elevate is an app that uses games to 'stay sharp, build confidence, and boost productivity – brain training personalised for you'; https://elevateapp.com, accessed 3 March 2023.

24. 'Care Home Fees and Costs: How Much do You Pay?', Carehome.co.uk, April 2022; https://www.carehome.co.uk/advice/care-home-fees-and-costs-how-much-do-you-pay, accessed 3 March 2023.

25. Aline Ollevier, Gabriel Aguiar, Marco Palomino and Ingeborg Sylvia Simpelaere, 'How Can Technology Support Ageing in Place in Healthy Older Adults? A Systematic Review', *Public Health Reviews*, no. 41, article no. 26 (23 November 2020); https://doi.org/10.1186/s40985-020-00143-4.

26. Alessandra Talamo, Marco Camilli, Loredana di Lucchio and Stefano Venture, 'Information from the Past: How Elderly People Orchestrate Presences, Memories and Technologies at Home', *Universal Access in the Information Society*, no. 16 (August 2017), pp. 739–53; https://doi.org/10.1007/s10209-016-0508-6, accessed 3 March 2023.

27. Frida Ryman, Jetske Erisman and Léa Darvey, 'Health Effects of the Relocation of Patients With Dementia: A Scoping Review to Inform Medical and Clinical Decision-Making', *The Gerontologist*, vol. 59, issue 6 (December 2019), pp. e674–e682; https://doi.org/10.1093/geront/gny031, accessed 3 March 2023.

28. Tori DeAngelis, 'Optimizing Tech for Older Adults', *Monitor on Psychology*, vol. 52, no. 5 (1 July 2021), p. 54; https://www.apa.org/monitor/2021/07/tech-older-adults, accessed 3 March 2023.

29. 'Lack of Sleep in Middle Age May Increase Dementia Risk', National Institute of Health; https://www.nih.gov/news-events/nih-research-matters/lack-sleep-middle-age-may-increase-dementia-risk, accessed 3 March 2023.

30. Eric Suni, 'Technology in the bedroom', Sleep Foundation, updated 15 December 2022; https://www.sleepfoundation.org/bedroom-environment/

technology-in-the-bedroom, accessed 3 March 2023.

31. Ann Fessler, *The Girls Who Went Away: The Hidden History of Women Who Surrendered Children for Adoption in the Decades Before Roe v Wade* (New York, Penguin Books, 2007).

32. Gregory Rodriguez, 'How Genealogy Became Almost as Popular as Porn', Time.com, 30 May 2014; https://time.com/133811/how-genealogy-became-almost-as-popular-as-porn/, accessed 3 March 2023.

33. Antonio Regalado, 'More than 26 Million People Have Taken an At-Home Ancestry Test', MIT Technology Review, 11 February 2019; https://www.technologyreview.com/2019/02/11/103446/more-than-26-million-people-have-taken-an-at-home-ancestry-test/, accessed 3 March 2023.

34. Heather Perlberg, 'Blackstone Reaches $4.7 Billion Deal to Buy Ancestry.com', Bloomberg.com, 5 August 2020; https://www.bloomberg.com/news/articles/2020-08-05/blackstone-said-to-reach-4-7-billion-deal-to-buy-ancestry-com?leadSource=uverify%20wall, accessed 3 March 2023.

35. Brody Ford and Emily Chang, 'Facebook Executive Deborah Liu Named CEO of Ancestry.com', Bloomberg UK, 3 February 2021; https://www.bloomberg.com/news/articles/2021-02-03/facebook-executive-deborah-liu-named-ceo-of-ancestry-com?leadSource=uverify%20wall, accessed 3 March 2023.

36. Diane Brady, 'Exclusive: Ancestry Expands DNA Testing to 54 New Markets', Forbes.com, 24 August 2022; https://www.forbes.com/sites/dianebrady/2022/08/24/exclusive-ancestry-expands-dna-testing-to-54-new-markets/?sh=5beffbc5c84c, accessed 3 March 2023.

37. Ibid.

38. Jacob Sonenshine, 'Where to Find Stocks that Will Rise 10 Times', Barrons.com, 8 November 2022; https://www.barrons.com/articles/stocks-can-rise-tenfold-consumer-tech-51667862981, accessed 3 March 2023.

39. Dani Shapiro, *Inheritance: A Memoir of Genealogy, Paternity, and Love* (London, Daunt Books, 2014).

40. Jacqueline Mroz, 'A Mother Learns the Identity of Her Child's Grandmother. A Sperm Bank Threatens to Sue', *New York Times*, 16 February 2019; https://www.nytimes.com/2019/02/16/health/sperm-donation-dna-testing.html, accessed 3 March 2023.

41. Personal communication with Jodi Klugman-Rabb, 29 March 2022.

42. https://righttoknow.us, accessed 3 March 2023.

43. Mroz, 'A Mother Learns the Identity of Her Child's Grandmother'.

44. Ibid.

45. Andrew Stern, 'The Hidden Dangers of DNA Tests: Do the Benefits Outweigh the Risks?', NBC/Think Again, 1 March 2019; https://www.nbcnews.com/think/video/home-dna-testing-is-your-privacy-at-risk-of-a-data-breach-1450470467980, accessed 3 March 2023.

46. https://righttoknow.us, accessed 3 March 2023.

47. Roger J. R. Levesque, 'Decisional Privacy', in Roger Levesque, *Adolescence, Privacy, and the Law: A Developmental Science Perspective* (Oxford, Oxford University Press, 2016), pp. 16–55; https://doi.org/10.1093/acprof:oso/9780190460792.003.0002, accessed 3 March 2023.

48. Kara Gavin, 'Older Adults Have High Interest in Genetic Testing – and Some Reservations', Michigan Medicine, 1 October 2018; https://www.michiganmedicine.org/health-lab/older-adults-have-high-interest-genetic-testing-and-some-reservations, accessed 3 March 2023.

49. Ibid.

50. Ibid.

51. Melissa Bailey, 'Genetic-testing Scan Targets Seniors and Rips off Medicare', NBCNews.com, 31 July 2019; https://www.nbcnews.com/health/aging/genetic-testing-scam-targets-seniors-rips-medicare-n1037186, accessed 3 March 2023.

52. https://www.ancestry.com, accessed 3 March 2023.

53. Stern, 'The Hidden Dangers of DNA Tests'.

54. Personal communication with Albert Gidari, 17 March 2021.

55. Alex Haley, *Roots* (New York, Vintage, 1994).

56. Alex Haley (writer/screenplay) and James Lee (screenplay), Marvin J. Chomsky, John Erman, David Greene and Gilbert Moses (directors), David L. Wolper (executive producer), *Roots: The Saga of an American Family* (TV miniseries), (Warner Bros. Television, 1977).

57. Sojourner Ahébée, 'For African Americans, DNA Tests Offer Some Answers Beyond the "Wall of Slavery"', Whyy.org, 21 May 2021; https://whyy.org/segments/tracing-your-ancestry-through-dna/, accessed 3 March 2023.

58. https://africanancestry.com/, accessed 3 March 2023.

59. Aaron Panofsky, 'Genetic Ancestry Testing Among White Nationalists: From Identity Repair to Citizen Science', *Social Studies of Science*, vol. 49, issue 5 (2 July 2019); https://doi.org/10.1177/0306312719861434, accessed 3 March 2023.

60. Brady, 'Exclusive: Ancestry Expands DNA Testing to 54 New Markets'.

61. Julia Creet (director, producer), *Data Mining the Deceased: Ancestry and the Business of Family* (2017); https://juliacreet.vhx.tv/products/data-mining-the-deceased-ancestry-the-business-of-family.

62. 'Baptism for the Dead', The Church of Jesus Christ of Latter-Day Saints, 2023; https://newsroom.churchofjesuschrist.org/article/baptism-for-the-dead, accessed 1 March 2023.

63. Personal communication with Professor Julia Creet, 20 March 2022.

64. https://www.legacytree.com/about, accessed 3 March 2023.

65. Paul Woodbury, 'Before It's Too Late: DNA Testing Older Relatives NOW', LegacyTree.com, 2017; https://www.legacytree.com/blog/dna-testing-older-relatives-now, accessed 3 March 2023.

66. https://www.juliacreet.com. My conversation with Professor Julia Creet was on 20 March 2022.

67. Julia Creet, *The Genealogical Sublime* (Amherst, MA, The University of Massachusetts Press, 2020).

68. Sigmund Freud, *The Unconscious* (London, Penguin Classics, 2005).

69. Heather Perlberg, 'Blackstone Reaches $4.7 Billion Deal to Buy Ancestry. com', Bloomberg.com, 5 August 2020; https://www.bloomberg.com/news/articles/2020-08-05/blackstone-said-to-reach-4-7-billion-deal-to-buy-ancestry-com?leadSource=uverify%20wall, accessed 3 March 2023.

70. Gina Kolata and Heather Murphy, 'The Golden State Killer is Tracked Through a Thicket of DNA, and Experts Shudder', *New York Times*, 27 April 2018; https://www.nytimes.com/2018/04/27/health/dna-privacy-golden-state-killer-genealogy.html, accessed 3 March 2023.

71. Nila Bala, 'We're Entering a New Phase in Law Enforcement's Use of Consumer Genetic Data', Slate, 19 December 2019; https://slate.com/technology/2019/12/gedmatch-verogen-genetic-genealogy-law-enforcement.html, accessed 3 March 2023.

72. Eric Rosenbaum, '5 Biggest Risks of Sharing Your DNA with Consumer Genetic-Testing Companies', CNBC Disruptor/50, 16 June 2018; https://www.cnbc.com/2018/06/16/5-biggest-risks-of-sharing-dna-with-consumer-genetic-testing-companies.html, accessed 3 March 2023.

73. Emily Mullin, 'Trump's DNA is Reportedly for Sale. Here's What Someone Could Do With It' [online article], OneZero/Medium.com, 14 February 2020; https://onezero.medium.com/trumps-dna-is-reportedly-for-sale-here-s-what-someone-could-do-with-it-e4402a9062c2, accessed 3 March 2023.

74. 'The Davos Collection: Predictive Artifacts of the Global Elite', Auction Catalog, EARNE$T, 2019; https://static1.squarespace.com/static/5c7e298c65a707492acb5b96/t/624c88f7f2c1692f6a3f09b2/1649182972785/The_Davos_Collection_LR.pdf, accessed 3 March 2023.

75. Personal communication with Albert Gidari, 17 March 2021.

76. Ibid.

77. 'Loneliness Research and Resources', AgeUK.org.uk, 8 February 2023; https://www.ageuk.org.uk/our-impact/policy-research/loneliness-research-and-resources/, accessed 24 Mar 2023.

Chapter 9: Digital Afterlife

1. https://remars.amazonevents.com, accessed 2 March 2023.

2. https://remars.amazonevents.com/learn/keynotes/?trk=www.google.com, accessed 2 March 2023. Rohit Prasad's 23 June 2022 talk starts at approximately 43:30.

3. See Elaine Kasket, *All the Ghosts in the Machine: The Digital Afterlife of Your Personal Data* (London, Robinson/Little Brown, 2020).

4. See, for example, James Vincent, 'Amazon shows off Alexa Feature that Mimics the Voices of Your Dead Relatives: Hey Alexa, That's Weird as Hell', TheVerge.com, 23 June 2022; https://www.theverge.com/2022/6/23/23179748/amazon-alexa-feature-mimic-voice-dead-relative-ai, accessed 1 March 2023.

5. The International Data Corporation (IDC) tracks and makes predictions about the Global StorageSphere: the size of the installed base of storage capacity, how much data is stored, and the amount of storage available every year: see https://www.idc.com/getdoc.jsp?containerId=IDC_P38353, accessed 2 March 2023. See also Sydney Sawaya, 'Stored Data Doubling Every 4 Years, IDC Says', SDX, 13 May 2020; https://www.sdxcentral.com/articles/news/stored-data-doubling-every-4-years-idc-says/2020/05/, accessed 2 March 2023.

6. 'Climate change puts energy security at risk', World Meteorological Organization, 11 October 2022; https://public.wmo.int/en/media/press-release/climate-change-puts-energy-security-risk, accessed 2 March 2023.

7. Nicola Jones, 'How to stop data centres from gobbling up the world's electricity', *Nature*, 12 September 2018; https://www.nature.com/articles/d41586-018-06610-y, accessed 2 March 2023.

8. 'How Much Energy Do Data Centers Really Use?', Energy Innovation Policy & Technology LLP, 17 March 2020; https://energyinnovation.org/2020/03/17/how-much-energy-do-data-centers-really-use/, accessed 2 March 2023.

9. Anna Castelnovo, Simone Cavallotti, Orsola Gambin and Armando D'Agostino, 'Post-bereavement hallucinatory experiences: A critical overview of population and clinical studies', *Journal of Affective Disorders*, 186 (1 November 2015), pp. 266–74; doi.org/10.1016/j.jad.2015.07.032; epub, 31 July 2015; PMID: 26254619. https://pubmed.ncbi.nlm.nih.gov/26254619/, accessed 2 March 2023.

10. Carl J. Öhman and David Watson, 'Are the Dead Taking Over Facebook? A Big Data Approach to the Future of Death Online', *Big Data and Society* (23 April 2019); https://doi.org/10.1177/2053951719984254, https://journals.sagepub.com/doi/full/10.1177/2053951719984254, accessed 2 March 2023.

11. 'Only 4 in 10 UK adults have a Will despite owning a property', Today's Wills and Probate, 13 August 2021; https://todayswillsandprobate.co.uk/only-4-in-10-uk-adults-have-a-will-despite-owning-a-property/, accessed 2 March 2023, and Rachel Lustbader, 'Caring.com's 2023 Wills Survey Finds that 1 in 4 Americans See a Greater Need for an Estate Plan Due to Inflation', Caring.com, 2023; https://www.caring.com/caregivers/estate-planning/wills-survey/, accessed 2 March 2023.

12. James Norris and Carla Sofka, 'The Digital Death Report 2018', Digital Legacy Association, 2018; https://digitallegacyassociation.org/wp-content/uploads/2019/11/Digital-Death-Survey-2018-The-Digital-Legacy-Association-1.pdf, accessed 2 March 2023.

13. Dave Lee, 'Twitter prepares for huge cull of inactive users', BBC.co.uk, 26 November 2019; https://www.bbc.co.uk/news/technology-50567751, accessed 2 March 2023.

14. Dave Lee, 'Twitter account deletions on "pause" after outcry', BBC.co.uk, 27 November 2019; https://www.bbc.co.uk/news/technology-50581287, accessed 2 March 2023.

15. 'Facebook Ruling: German Court Grants Parents Right to Dead Daughter's Account', BBC.co.uk, 12 July 2018; https://www.bbc.co.uk/news/world-europe-44804599, accessed 3 March 2023.

16. Rosa Marchitelli, 'Apple Blocks Widow from Honouring Husband's Dying Wish', CBC News, 19 October 2020; https://www.cbc.ca/news/business/widow-apple-denied-last-words-1.5761926, accessed 3 March 2023.

17. BBC, 'Facebook Removes Hollie Gazzard Photos with her Killer', BBC.co.uk, 11 November 2015; https://www.bbc.co.uk/news/uk-england-gloucestershire-34781905, accessed 3 March 2023.

18. See Kasket, *All the Ghosts in the Machine*, pp. 93–4.

19. Ben Stevens, 'Widow Wins Long Battle for iPhone Family Photos', *The Times*, 11 May 2019; https://www.thetimes.co.uk/article/widow-wins-long-battle-for-iphone-family-photos-h7mv9bw7t, accessed 3 March 2023.

20. Norris and Sofka, 'The Digital Death Report 2018'.

21. Google's Inactive Account Manager allows you to nominate, within the platform, someone you would like to have access to your Google accounts after your confirmed death. For an explanation of how it works, see https://digitallegacyassociation.org/google-guide/, accessed 3 March 2023.

22. For a tutorial on managing digital assets on Facebook and a tutorial on Digital Legacy, see https://digitallegacyassociation.org/facebook-tutorial/, accessed 3 March 2023.

23. Lernert Engelberts and Sander Plug, 'I Love Alaska' (Netherlands, Submarine Channel, 2009). This documentary film chronicles the search history of user 711391 from the 2006 search-data leak from America Online (AOL). See also Nat Anderson, '2006 AOL search Data Snafu Spawns

"I Love Alaska" Short Films', ArsTechnica, 28 January 2009; https://arstechnica.com/information-technology/2009/01/aol-search-data-spawns-i-love-alaska-short-films/, accessed 3 March 2023.

24. Personal communication with Al Gidari, 15 January 2021. Information about Al Gidari can be found on the website for the Center for Internet and Society at Stanford Law School; https://cyberlaw.stanford.edu/about/people/albert-gidari, accessed 3 March 2023.

25. Kasket, *All the Ghosts in the Machine*, pp. 102–8.

26. Charlie Brooker (writer) and Owen Harris (director), 'Be Right Back' (episode of *Black Mirror*), Channel 4, 11 February 2013.

27. Debra Bassett, 'Who Wants to Live Forever? Living, Dying and Grieving in Our Digital Society', *Social Sciences*, vol. 4, issue 4 (20 November 2015), pp. 1127–39; https://doi.org/10.3390/socsci4041127, available on https://www.mdpi.com/2076-0760/4/4/1127, accessed 3 March 2023.

28. Vanessa Nicolson, 'My Daughter's Death Made Me Do Something Terrible on Facebook', *Guardian*, 22 April 2017; https://www.theguardian.com/lifeandstyle/2017/apr/22/overcome-grief-daughter-death-died-message-boyfriend-facebook?, accessed 3 March 2023.

29. Dhruti Shah, 'Somebody Answered My Dead Brother's Number', BBC.co.uk, 10 October 2019; https://www.bbc.co.uk/news/world-us-canada-49941840, accessed 3 March 2023.

30. BBC, 'Kanye West Gives Kim Kardashian Birthday Hologram of Dead Father', BBC.co.uk, 30 October 2020; https://www.bbc.co.uk/news/entertainment-arts-54731382, accessed 3 March 2023.

31. Olivia Harrison, 'Wait, How Much Did Kim Kardashian West's Birthday Hologram Cost?', Refinery29.com, 31 October 2020; https://www.refinery29.com/en-gb/2020/10/10140853/kim-kardashian-dad-hologram-cost, accessed 3 March 2020. See also Heather Schwedel, 'How You Create a Robert Kardashian-Style Hologram – and How Much it Costs', Slate.com,

31 October 2020; https://slate.com/human-interest/2020/10/robert-kardasian-hologram-company-kim-kanye-cost.html, accessed 3 March 2023.

32. David Rowell, 'The Spectacular, Strange Rise of Music Holograms', *Washington Post*, 30 October 2019; https://www.washingtonpost.com/magazine/2019/10/30/dead-musicians-are-taking-stage-again-hologram-form-is-this-kind-encore-we-really-want/, accessed 3 March 2023.

33. Lane Brown, 'We've Been Thinking About Holograms All Wrong: Forget Reanimating Dead Musicians. This Tech is for Living Performers Who Can't Stand Their Bandmates', *New York Magazine: Vulture*, 11 October 2022; https://www.vulture.com/2022/10/abba-voyage-london-holograms.html, accessed 3 March 2023.

34. Morgan Neville (director), Eileen Myers and Aaron Wickenden (editors), *Roadrunner: A Film About Anthony Bourdain*, CNN Films/HBO Max/Tremolo Productions/Zero Point Zero (2021).

35. Helen Rosner, 'The Ethics of a Deepfake Anthony Bourdain Voice', *The New Yorker*, 17 July 2021; https://www.newyorker.com/culture/annals-of-gastronomy/the-ethics-of-a-deepfake-anthony-bourdain-voice, accessed 3 March 2023.

36. Responding to an article published in *Variety*, Ottavia Busia posted on 16 July 2021; https://twitter.com/OttaviaBourdain/status/1415889455005716485?lang=en, accessed 3 March 2023.

37. Anthony Bourdain was known for *Parts Unknown*, a travel and food show that ran for 12 seasons on CNN.

38. See, for example, Alex Marshall, 'Abba Returns to the Stage in London. Sort Of', *New York Times*, 27 May 2022; https://www.nytimes.com/2022/05/27/arts/music/abba-voyage-london.html, accessed 3 March 2023.

39. Cathy Scott, 'Digital Domain Cashes In On "Hologram Tupac"', Forbes, 10 May 2012; https://www.forbes.com/sites/crime/2012/05/10/digital-domain-cashes-in-on-hologram-tupac/?sh=603c4fb27d45m, accessed 3 March 2023.

40. See https://remars.amazonevents.com/learn/keynotes/?trk=www.google.com.

41. Descript advertises 'Ultra-realistic voice cloning with Overdub. Descript's Overdub lets you create a text-to-speech model of your voice or select from one of our ultra-realistic stock voices.' See https://www.descript.com, accessed 3 March 2023.

42. Nilay Patel, Eleanor Donovan, Josh Barry, Chad Mumm, Mark Olsen, Max Heckman, John Warren and Chris Gross (executive producers), *The Future of Life After Death* (episode of *The Future of*), The Verge/21 Laps Entertainment/Vox Media Studios, Netflix, 28 June 2022.

43. Cole Imperi is a thanatologist, author and the founder of the School of American Thanatology; https://coleimperi.com, accessed 3 March 2023.

44. Matthew Groh, Ziv Epstein, Chaz Firestone and Rosalind Picard, 'Deepfake Detection by Human Crowds, Machines, and Machine-Informed Crowds', *PNAS*, vol. 119, no. 1 (5 January 2022); https://doi.org/10.1073/pnas.2110013119, accessed 3 March 2023.

45. As time goes on, we will doubtless see numerous and notable examples of the dead being spoken for/made to serve the living, with various consequences. For example, the digital dead have been posthumously tied to the whipping post and trolled for political reasons. See for example Dan Levin, 'They Died from Covid. Then the Online Attacks Started', *New York Times*, 27 November 2021; https://www.nytimes.com/2021/11/27/style/anti-vaccine-deaths-social-media.html, accessed 3 March 2023.

46. Dalvin Brown, 'AI Chat Bots Can Bring You Back from the Dead, Sort Of', *Washington Post*, 4 February 2021; https://www.washingtonpost.com/technology/2021/02/04/chat-bots-reincarnation-dead/, accessed 3 March 2023.

47. On 22 January 2021, Tim O'Brien posted, 'At any rate, confirmed that there's no plan for this. But if I ever get a job writing for Black Mirror, I'll know to go to the USPTO website for story ideas'; https://twitter.com/_TimOBrien/status/1352674749277630464?s=20, accessed 3 March 2023.

48. Jason Fagone, 'The Jessica Simulation: Loss and Loss in the Age of A.I.', *San Francisco Chronicle*, 23 July 2021; https://www.sfchronicle.com/projects/2021/jessica-simulation-artificial-intelligence, accessed 3 March 2023.

49. The Project December service advertises itself explicitly as being able to simulate the dead; https://projectdecember.net, accessed 3 March 2023.

50. Fagone, 'The Jessica Simulation'.

51. Storyfile advertises itself as a service for 'Conversational video – videos that talk back. We make AI feel more human'; https://storyfile.com, accessed 3 March 2023.

52. Joacquin Victor Tacia, 'An "AI-Powered Hologram" Creepily Revives Dead Woman to Deliver A Speech at Her Funeral', TechTimes, 18 August 2022; https://www.techtimes.com/articles/279328/20220818/ai-ai-powered-hologram-creepily-revives-dead-woman-deliver-speech-funeral.htm, accessed 3 March 2023.

53. William Shatner has been reported as joining with Storyfile to 'explore the new frontier of AI'. See https://storyfile.com/william-shatner-joins-new-frontier-of-ai-at-90/, accessed 3 March.

54. Kim Jong-Woo (director), *Meeting You* (MBC America, 2020). See Lee Gyu-Lee, '"Meeting You" Creator on His Controversial Show: "I Hope it Opens Up Dialogue"', *The Korea Times*, 5 April 2020; https://www.koreatimes.co.kr/www/art/2020/04/688_287372.html, accessed 3 March 2023.

55. There are often reactions and judgements about other people's grief, including (and perhaps especially) grieving and mourning involving technology. It's important to note that there is no one grieving process, and that in fact the digital environment has in some ways revived unhelpful notions about stage-based grieving and normative grieving. See Morna O'Connor and Elaine Kasket, 'What Grief Isn't: Dead Grief Concepts and their Digital-Age Revival', in T. Machin, C. Brownlow, A. Abel and

J. Gilmour (eds), *Social Media and Technology Across the Life Span* (*Palgrave Studies in Cyberpsychology*), (Cham, Switzerland, Palgrave Pivot, 2022), pp. 115–30.

56. Edwina Harbinja, Lilian Edwards and Marisa McVey, 'Governing Ghostbots', *Computer Law & Security Review*, vol. 48 (April 2023); https://doi.org/10.1016/j.clsr.2023.105791, accessed 3 March 2023.

57. Michael Hvid Jacobsen (ed.), *The Age of Spectacular Death* (London, Routledge, 2020).

58. At the time of writing, John Troyer was associated with the Centre for Death and Society at the University of Bath. His TEDx talk is available on https://www.youtube.com/watch?v=UjIrVfqKWLQ, accessed 2 March 2023.

59. Patrick Stokes, *Digital Souls: A Philosophy of Online Death* (London, Bloomsbury Publishing, 2021).

Conclusion

1. This is a riff on the Serenity Prayer, attributed to American theologian Reinhold Niebuhr in 1943.

2. Stephen Covey, *The Seven Habits of Highly Effective People: Powerful Lessons in Personal Change* (reissue edn) (London, Simon & Schuster, 2013).

3. For links to a collection of values exercises, see Elaine Kasket, 'Goals, Values, and Technology' (*This is Your Life on Tech* podcast episode); https://lifeontech.substack.com/p/goals-values-and-technology-ea7 (23 January 2023), accessed 8 March 2023.

4. Jenny Odell, *How to do Nothing: Resisting the Attention Economy* (London, Melville House Publishing, 2021).

5. John Tierney and Roy F. Baumeister, *The Power of Bad: How the Negativity Effect Rules Us and How We Can Rule It* (New York: Penguin, 2019).

6. Martin Heidegger, *Being and Time* (New York, HarperCollins, 2008).

Index

About the author

As a hybrid Counselling Psychologist and Cyberpsychologist, Dr Elaine Kasket is a uniquely qualified guide to the challenges and opportunities of the modern world. Through her writing, keynote speaking and media appearances, she helps people deepen their wellbeing, maximise their connections and retain their fundamental humanity in the digital age. She is the author of *All the Ghosts in the Machine: The Digital Afterlife of Your Personal Data* (2019); an Honorary Professor of Psychology at the University of Wolverhampton; a Fellow of the Royal Society of Arts; and an Associate Fellow of the British Psychological Society. American by birth, she is based in London, where she runs a busy coaching and psychotherapy practice.